**LIFE CYCLE COSTING:
A PRACTICAL GUIDE FOR
ENERGY MANAGERS**

LIFE CYCLE COSTING: A PRACTICAL GUIDE FOR ENERGY MANAGERS

Robert J. Brown
Capitol Campus,
The Pennsylvania State University

Rudolph R. Yanuck, P.E.
Partner
EER Associates

THE FAIRMONT PRESS, INC.
P.O. BOX 14227
ATLANTA, GEORGIA 30324

Life Cycle Costing: A Practical Guide for Energy Managers

© 1980 by The Fairmont Press

All rights reserved. Printed in the United States of America. No part of this publication may be reproduced, stored in a retrieval system, or transmitted in any form or by any means, electronic, mechanical, photocopying, recording or otherwise, without the prior written permission of the publisher.

While every effort is made to provide dependable information, the publishers and authors cannot be held responsible for any inaccuracy.

ISBN: 0-915586-17-7

...... for having shared us with a demanding undertaking and for their help and encouragement in its completion, we dedicate this book to our wives .....
Irene S. Brown   *Carolyn J. Yanuck*

Foreword

The purpose of this book is to explain the concept of life cycle costing, to suggest the kinds of purchases for which it may be used to advantage, and to provide users with a handy manual of techniques and applications of life cycle costing. Lengthy descriptions are minimized in favor of an abundance of real-life examples which are solved. The mathematics rarely goes beyond elementary arithmetic and in those few areas where it does, the material is optional and to skip over it would in no way interfere with the reader's comprehension of the text. A bibliography is provided so that those who seek more information on particular subjects will know where to find it.

Energy conservation receives considerable attention since the escalation of fuel prices has made energy costs an important consideration in the purchase of a wide variety of products. Although the technical terminology pertinent to each illustration is explained in the problems, a special chapter is provided to give the reader a broader background on energy fundamentals. This material will facilitate application of life cycle costing to a broad spectrum of energy-related projects.

Special features in the text include: a concise summary of tax legislation pertaining to write-offs of energy-conserving expenditures, a section on ranking of projects (including Department of Energy requirements), a set of discount-escalation tables developed by the authors and not widely available, a simple explanation of a computer program having life cycle costing application, and a group of case problems. Solutions to all problems in the text are provided.

This book has been developed by the authors from materials used in conducting workshops on life cycle costing and many of the problems and illustrations are based upon their consulting experience. The book has been written for the benefit of the user and should be of value to a broad spectrum of readers: to those with purchasing responsibilities in both private and public sectors, to dealers and manufacturers, to engineers, architects, and consultants. It is the hope of the authors that the concepts and methodologies expounded in the book will find wide application, and that the benefits derived by the users will reflect significant progress in the optimization of the nation's scarce resources.

Contents

Foreword vii

1 Concept of Life Cycle Costing........................1
   Objective • Current Usage of LCC • When to Use LCC • LCC: Its Time Has Come • LCC: Opportunities

2 The Time Value of Money............................12
   Terminology • Time Value of Money • Frequency of Compounding • Continuous Compounding • Continuous Cash Flows • Self-Study Problems

3 Application of Life Cycle Costing...................37
   Present Worth Method • Uniform Annual Cost Method • Forever Projects • Self-Study Problems

4 Costs...49
   Classification • Interest Cost • Depreciation • Format • Cost Determination • Information Sources • Cost Estimating • Sensitivity Analysis • Expected Value Analysis • Self-Study Problems

5 Energy Cost Estimation...............................67
   Abbreviations • Important Energy Values • Formulas for Computing Building Energy Costs • Electrical Power Cost • Cost Reduction Ideas

6 Service Life..93
   Planning Period • Determining Service Life • Self-Study Problems

Contents

7 Taxes and Depreciation 99
Tax Considerations • Depreciation •
Accelerated Depreciation • ADR System •
After-Tax Costs • Self-Study Problems

8 Lease or Buy 107
Types of Leases • Leveraged Leases •
Lease or Buy • Discount Rate •
Self-Study Problems

9 Replacements, Life Differences 112
Replacement • Tax Effects • Life Differences:
Chain Method • Self-Study Problems

10 Escalation 119
Car Problem • Escalating Costs: Present Worth •
Discount-Escalation Tables • Interval Costs •
Escalating Costs: Average Annual Cost •
Self-Study Problems

11 Payback, Break-Even Analysis 130
Payback • Simple Payback • Discounted
Payback • Logarithm Method • True Payback •
Break-Even Analysis • Self-Study Problems

12 Ranking 141
Net Present Worth (Value) Method •
Savings/Investment Ratio Method •
Btu Method • Required Ranking • Multiple
Projects • Escalation • Department of
Energy Method • Self-Study Problems

13 Computer Analysis 150
Methodology • Sensitivity Analysis • LCPBM •
Self-Study Problems

14 Discount Rates 157
Rates Matter • Private Sector • Weighted
Average Cost of Capital • Marginal Cost of
Capital • Risk Adjustment • Investment Rates •
Public Sector • Federal Government •
State and Local Governments • Self-Study
Problems

Case Studies 167
Appendices 187
Glossary 203
Bibliography 211
Time Value of Money Formulas 214
Interest Tables. 217
Discount–Escalation Factors 247
Solutions. 269
Index. 297

1

Concept of Life Cycle Costing

When a consumer decides that a certain item in a store is very desirable, the factor which is usually primary in the purchase decision is price. Subsequent costs which can be expected play a minor role. Whether the item be a piece of clothing, a television set, or even a house, anticipated maintenance costs rarely are a significant consideration in the decision-making process. Yet, maintenance costs are a price of ownership just as well as initial costs.

Life cycle costing (LCC) is a method of calculating the total cost of ownership over the life span of the asset. Initial cost and all subsequent expected costs of significance are included in the calculations as well as disposal value and any other quantifiable benefits to be derived. The LCC technique is justified whenever a decision must be made on the acquisition of an asset which will require substantial operating and maintenance costs over its life span.

"Lowest bid" as a decision criterion for construction of a building, as an example, makes no sense if the lowest bid is defined as initial cost. Operating and maintenance costs over the long life of a building far exceed initial costs and must be factored into the decision process. It may be more economically feasible to pay a higher initial cost in order to obtain a lower total ownership cost. Where expected life-time costs are high relative to initial cost "lowest bid" should be based on life cycle costs.

OBJECTIVE

The objective of this book is to explain the concept of life cycle costing and to show how the analysis is performed and how it contributes to cost-effective decision processes. A wide variety of real-life problems is used to demonstrate the principles.

CURRENT USAGE OF LCC

Life cycle costing has a long history of use in industry. Included within the framework of capital budgeting, LCC has been applied to an endless variety of projects and has been a decision-making aid which has played an important role in the success of many of the profit-oriented elements of our free enterprise system.

LCC analysis has had a long tradition in the U.S. Department of Defense and it is applied to virtually every new weapon system proposed or under development. The impact on the defense and aerospace industries has been so great that those industries now design their products in terms of LCC objectives. This practice is referred to as "Design to Cost."

Interest is now being shown in the use of LCC in the health care field. A General Accounting Office study in 1972 which revealed that the operating costs of a hospital in its first three to five years of existence typically exceeded the entire cost of construction did much to stimulate interest in cost-effective techniques. As a result of the report, the Department of Health, Education and Welfare initiated a project to formalize a life cycle costing model for use in the health field.

Surprisingly, the building industry has been slow to adopt LCC. Spurred on, however, by escalating operating costs and the prompting of the U.S. General Services Administration and the Department of Energy, builders are now becoming quite conscious of the advantages of LCC. Data accumulation and organization have been a problem but GSA and the American Institute of Architects have cooperated in the development of a costing framework—entitled UNIFORMAT—which should be useful for both public and private work. (The UNIFORMAT outline is provided in Appendix B.)

States have begun to use LCC for a variety of purposes, and in the past five years Florida, Alaska, North Carolina, New Mexico, Texas, Washington, California, and Maryland have passed legislation mandating LCC analysis for public building projects. (See Figure 1-1.) The State Energy Conservation Program, which was established by the Energy Policy and Conservation Act (ECPA) of 1975, has generated further interest and almost every state has submitted an LCC plan to qualify for the funds authorized by the program. Further, these plans outline training programs being developed to encourage the use of LCC for energy conservation in products purchased.

STATE	DATE	BILL
FLORIDA	1974	FECBA
ALASKA	1975	H.B. 429
NORTH CAROLINA	1975	S.B. 151
NEW MEXICO	1975	H.B. 395
WASHINGTON	1975	S.B. 2106
TEXAS	1976	S.B. 516
MARYLAND	1978	CHAPT. 596
CALIFORNIA	1978	TITLE 24

Figure 1-1. States Legislating LCC

In the years ahead, the move toward LCC is certain to gain greater impetus. Post-purchase costs of labor, materials, and energy are already important components of life cycle cost, and are likely to grow, relative to initial cost, as long as inflation resists control efforts and dependence on scarce energy resources continues. Life cycle costing makes good economic sense and its adoption can contribute to the conservation of our precious fuel supplies.

WHEN TO USE LCC

Life cycle cost analysis should not be used in each and every purchase. The process itself carries a cost and therefore can add to the cost of the commodity. LCC analysis can be justified only in those cases in which the cost of the analysis can be more than offset by the savings derived through the purchase of the commodity.

Four major factors which may influence the economic feasibility of applying LCC analysis are:

1. *Energy Intensiveness.*—LCC should be considered when the anticipated energy costs of the purchase are expected to be large throughout its life.

2. *Life Expectancy.*—For commodities with long lives, costs other than purchase price take on added importance. For commodities with short lives, the initial costs become a more important factor.

3. *Efficiency.*—The efficiency of operation and maintenance can have significant impact on overall costs. LCC is beneficial when savings can be achieved through reduction of maintenance costs.

4. *Investment Cost.*—As a general rule, the larger the investment the more important LCC analysis becomes.

An obvious example of a purchase which would satisfy all four of these criteria is a building. Buildings are heavy users of energy, they have long lives, they require maintenance and repair, and they are large investments. Indeed, there is abundant evidence that the post-purchase costs of buildings far exceed initial costs.

Some other assets for which LCC would be appropriate are: construction equipment, pollution control equipment, transportation vehicles, heating, ventilating and air-conditioning systems, farm equipment, and hospital equipment.

The four major factors listed above are not, however, necessary ingredients for life cycle cost analysis. An energy-efficient light bulb, for example, is not a large investment and its life is relatively short. Yet it is a proper subject for LCC analysis. A quick test to determine whether life cycle costing would apply to a purchase is to ask whether there are any post-purchase costs associated with it. Paper clips, pencils and safety pins are commodities which would not pass the test. Lowest initial cost would be the appropriate criterion for such items. A product such as a dining room table would have post-purchase maintenance costs, but unless the difference between those costs and the maintenance costs of a competing table were expected to be significant, LCC would not be worthwhile.

Laundry detergents are a commodity which can illustrate the difference between what is and what is not life cycle costing. If a large consumer of detergents invites bids to meet detailed specifications for soil removal, whiteness retention and discoloration levels of treated fabrics, inhibition of bacterial activity, etc., the choice should be determined on the basis of lowest "use" cost, that is, the lowest cost to wash a specified amount of laundry. This is an example of cost-effectiveness, but life cycle costing is not involved since the "use" cost here is an initial cost only.

On the other hand, if the buyer can control the temperatures of the washing machines and wants to compare bids not only in terms of "use" cost but also the subsequent differences in energy costs associated with the differences in minimum temperatures at which the various detergents are effective, life cycle costing may be used. The key element is the existence of post-purchase costs. Life cycle costs are a combination of initial and post-purchase costs.

LCC: ITS TIME HAS COME

It is unfortunate that the merits of life cycle costing took so long to be recognized. But prior to the 1970s the impetus was not very great. For one thing, fuel costs were a small fraction of total expected costs since resources were considered sufficient to provide unlimited energy at low cost.

Inflation in the 1950s and 1960s had generally been in the 1- to 2-percent range. However, in the 1970s, annual inflation rates in the 5- to 9-percent range (see Figure 1-2) were experienced. As a result, expected operating and maintenance costs are increasing in relative importance.

Failure to understand the concept and methodology of LCC has been another factor in its tardy acceptance by the public. Even today, although massive efforts are being made to foster a consciousness of and the use of life cycle costing, the term is scarcely a familiar one to the average citizen.

Finally, even many of those who have an appreciation for LCC are reluctant to use it because of the difficulty of estimating future costs. Calculations based on forecasts are widely con-

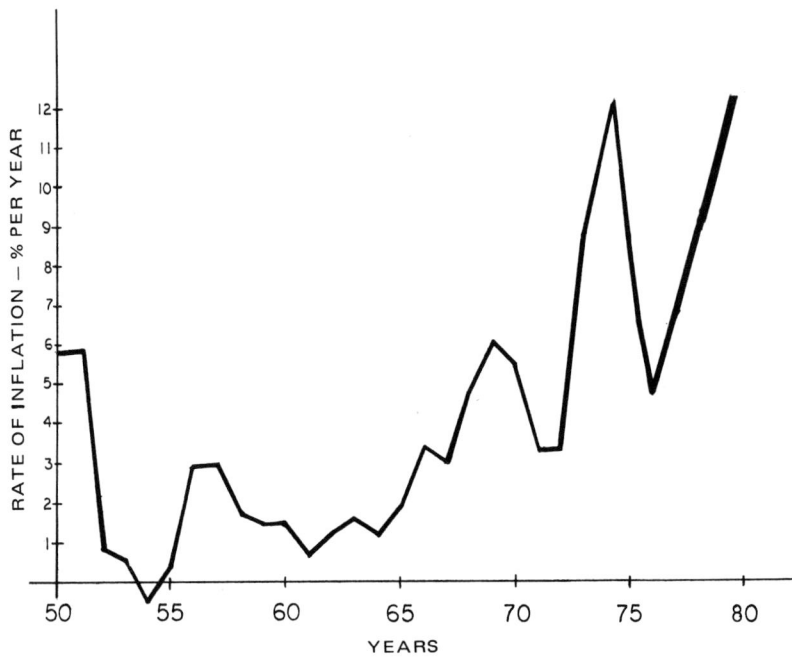

Figure 1-2. U.S. Consumer Price Index
(Source: Bureau of Labor Statistics, U.S. Department of Labor)

sidered to have dubious value. (Realization that a failure to forecast costs is the equivalent of attributing a zero value to them has not been an antidote to this skepticism.)

Traditions do not yield easily, and certainly the long tradition of emphasis on lowest initial cost has not been an exception. Entrenched ideas are not easily displaced by new methods such as LCC. And even today, after so much has been said and written on LCC, vast amounts of public and private purchasing is still based on lowest initial cost.

LCC: OPPORTUNITIES

The importance of including post-purchase costs in the evaluation of proposed expenditures can not be overemphasized. Informed buyers are already aware of this. Still, the potential savings of resources that can be achieved by a broader understanding of life cycle costing is enormous.

Concept of Life Cycle Costing 7

The scarcity and price escalation of fuel have contributed forcefully to the economic education of the public, but the less-dramatized costs and shortages, even yet, receive little attention in purchase decisions. Few home buyers realize that, with no inflation, over a 30-year period of ownership the annual costs of a house will amount to approximately double the initial cost (Figure 1-3). With an average 9% annual inflation rate the annual costs will be nine times the purchase price (Figure 1-4). The cost of electricity, scarcely considered in the purchase of appliances, will exceed the initial cost for such items as refrigerators, ranges, and clothes dryers over their lifetimes. Consumer education on the merits of LCC, especially for purchases requiring heavy use of energy, will not only improve living standards but will also stretch resources.

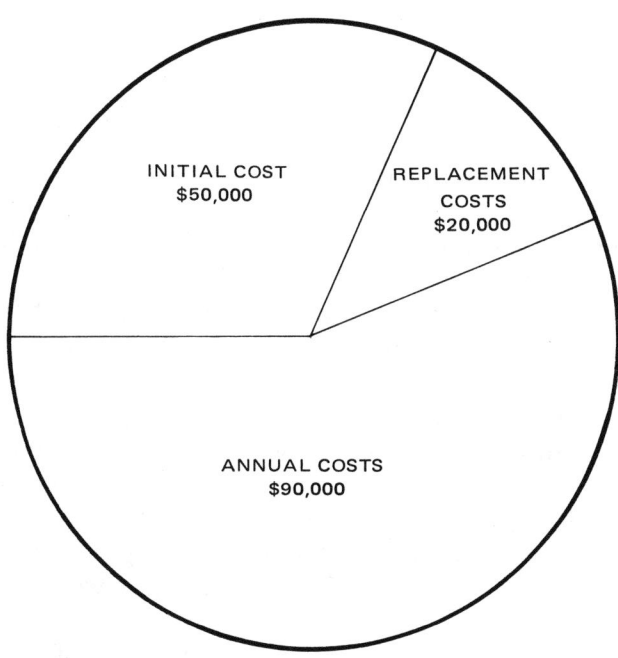

Figure 1-3. Owning a House—30 Years (With No Inflation)

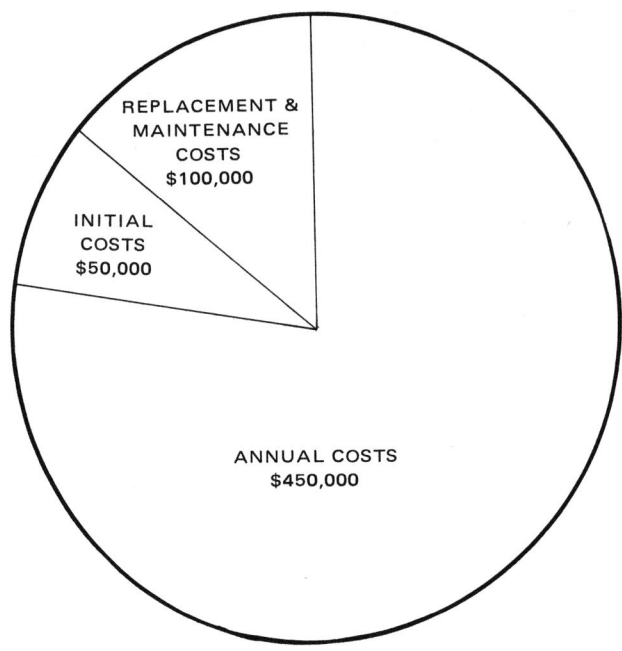

Figure 1-4. Owning a House—30 Years (9% Inflation Rate)

Myriad opportunities for savings may be found in the construction of all types of buildings and in retrofits of existing structures. Federal legislation, especially the Energy Act of 1978, has been particularly supportive of energy-conserving outlays in buildings. Life cycle costing is the tool for determining which outlays are economically justified.

Opportunities abound throughout the private and public sectors for effective use of LCC. The intent here is not to provide an exhaustive listing of these opportunities, but the sizable sampling of applications found in the following chapters should certainly sensitize the reader to circumstances where benefits may be achieved. Recognition of opportunities for savings is a giant step forward in the efficient management of resources. Throughout the text several examples are given with the solutions worked out. Included are problems you may wish to solve. Answers to these problems are included in the appendices.

Concept of Life Cycle Costing

Example 1-1

It seems appropriate in this introductory chapter to provide an illustration of a common purchase susceptible to analysis by life cycle costing. A first step is taken here in the development of the LCC process. This problem is used in later chapters progressively embellised to illustrate methodology. The elegance of the technique should be fully apparent in the final solution.

A familiar problem to the consumer is the purchase of a car. Features such as style, size, power, handling, comfort, etc. are considered in the mix with price, miles per gallon, and maintenance expectations, and a decision is made. If all the costs and expected costs were considered separately, however, and the LCC calculated, the decision could be made on the basis of whether the features of a particular car are worth the cost differentials of the alternative(s).

Let us take as an example a case where a buyer, having sufficient funds to buy any one of three desirable cars, decides to utilize LCC as a decision aid. Alternative investment opportunities will be ignored at this time. After some research, the buyer has drawn up the following table of information.

	CAR A	CAR B	CAR C
Purchase Price	$3,200	$ 3,800	$ 3,910
Sales Tax	5%	5%	5%
Salvage Value	100	130	200
License/Year	24	24	24
Miles/Gallon	18	24	28
Miles Between Tune-Ups	8,000	12,000	15,000
Insurance/Year	191	221	327

Other Information: Tune-ups cost $80 each, owner trades in every four years, gasoline is $.60 a gallon, the user drives 22,000 miles a year.

Assume all money spent during a given year is spent at the end of the year.

SOLUTION (Supporting calculations follow)

	CAR A	CAR B	CAR C
Purchase Price	$3,200	$3,800	$3,910
Sales Tax	160	190	195
Initial Cost	3,360	3,990	4,105
Differentials		630	745

By initial cost method, the cheapest is Car A and the differentials for B and C are $630 and $745 respectively. However, let us continue:

Annual Costs	CAR A	CAR B	CAR C
Fuel	$ 733	$ 550	$ 471
Maintenance	200	140	100
License	24	24	24
Insurance	191	221	327
	$1,148	$ 935	$ 922
Total Annual Costs, 4 years	$4,592	$3,740	$3,688
Add Initial Cost	3,360	3,990	4,105
Total	$7,952	$7,730	$7,793
Deduct Trade-In	(100)	(130)	(200)
Final Cost	$7,852	$7,600	$7,593
Differentials	$ 259	$ 7	--

By including life span costs, we find that C has the lowest total cost.

SUPPORTING CALCULATIONS

Fuel
A. 22,000 ÷ 18 = 1,222 gallons
    1,222 gallons × .60 = $733
B. 22,000 ÷ 24 = 917 gallons
    917 gallons × .60 = $550
C. 22,000 ÷ 28 = 786 gallons
    786 gallons × .60 = $471

Maintenance
A. 22,000 × 4   = 88,000 miles total in four years
    88,000 ÷ 8,000   = 11 tune-ups (but use 10 because the 11th tune-up occurs at end of life).
    10 × $80   = $800
    800 ÷ 4 years   = $200 per year
B. 88,000 ÷ 12,000 = 7 tune-ups
    7 × $80   = $560
    560 ÷ 4 years   = $140 per year
C. 88,000 ÷ 15,000 = 5 tune-ups

5 × $80 = $400
$400 ÷ 4 years = $100 per year

The solution to the car problem would be correct except for one very important consideration. Money has a time value. Regardless of whether the car is financed with a loan or purchased outright, the time value of the money must not be omitted. *Moreover, it would be incorrect to merely add the interest cost of borrowed money to the annual costs of a purchase.* There are well-established procedures for including the time value of money in life cycle cost analysis and these should be followed.

The procedures for dealing with the time value of money are explained in chapter 2.

2

The Time Value of Money

TERMINOLOGY

The term "capital budgeting" is defined as the process of analyzing expenditures on assets (expected to provide returns beyond a year) to determine if they should be included in the capital budget. The outlays may be for projects which are either revenue-producing or nonrevenue-producing.

Life cycle costing is not something different from capital budgeting but, rather, is generally understood to be the application of capital budgeting to nonrevenue-producing projects. Capital budgeting is a term more familiar to industry while life cycle costing is more familiar to government. The principles are the same, however, regardless of the terminology since the objective is to maximize benefits and minimize costs. For nonrevenue-producing projects such questions arise as:

"Will it pay for itself?"
"If so, in what period of time?"
"Which of several alternatives will cost the least or save the most?"
"Should it be leased or bought?"
"Should it be replaced?"

These are questions which can be answered by the application of life cycle costing methods.

Maximizing savings and minimizing costs is an objective of both the public and the private sectors. Even though the former is motivated by social benefit and the latter by profit, both have the same analytical goals and these goals are achieved by optimal productivity of capital.

Capital has a cost, regardless of whether it comes from an investor voluntarily or from a taxpayer involuntarily. A business derives its capital from retained profits, borrowed funds, and sale of shares in the business. The overall cost of its financing is referred to as the "cost of capital." Government obtains funds primarily from taxation and sale of bonds, and although the "cost of capital" definition is not exactly parallel, the application to capital analysis is similar.

The expression "opportunity cost" is occasionally encountered in LCC analysis. It refers to the cost sacrificed by not investing in an alternative project. For example, a person who is contemplating investing $30,000 in a franchise should include in the analysis the opportunity cost of, say, 8% which could have been earned in an alternative investment. As long as capital can be employed in other projects and earn a return, it is not free—it has an opportunity cost. This applies even where the capital takes the form of assets other than cash as long as a cash-equivalent value can be established for the assets.

TIME VALUE OF MONEY

To say that capital has an opportunity cost is another way of saying that money has a time value. If $100 can be invested today at a 6% annual rate it will be worth $100 \times 1.06 = \$106$ a year from now. If the investment continues for a second year it will be worth $100 \times 1.06^2 = \$112.36$ (or 106×1.06). This process is referred to as compounding. An investment of $100 compounded for 5 years at 6% will be worth $100 \times 1.06^5 = \$133.82$ (see Figure 2-1).

The present worth of an amount of money due in the future is calculated by a process known as discounting. In the case of the above illustration, the question might be asked, "If I can earn 6% on a savings account, how much would I have to invest today in order to have $133.82 in 5 years?" The answer would be $133.82 \div 1.06^5 = \$100$ (see Figure 2-2). In other words, discounting is the reciprocal of compounding.

The discounting process is particularly important in life cycle cost analysis because it facilitates the translation of future values to present values. If the total cost of owning an asset is

14 Life Cycle Costing: A Practical Guide for Energy Managers

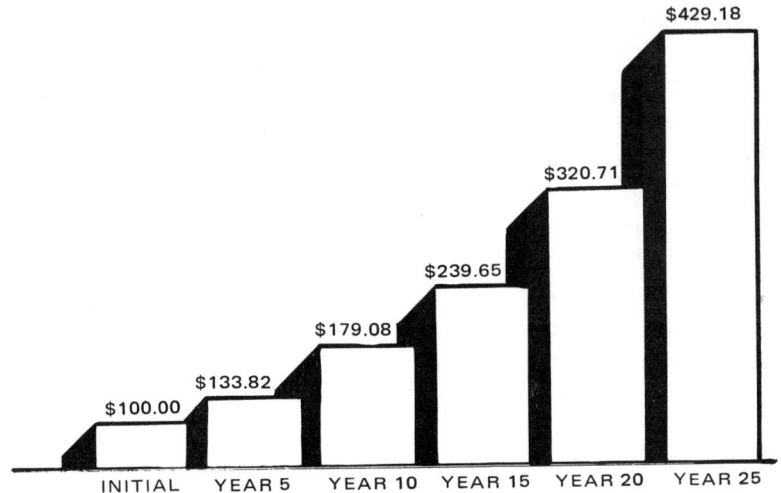

Figure 2-1. Time Value of Money: The Compounding Effect
i = 6%

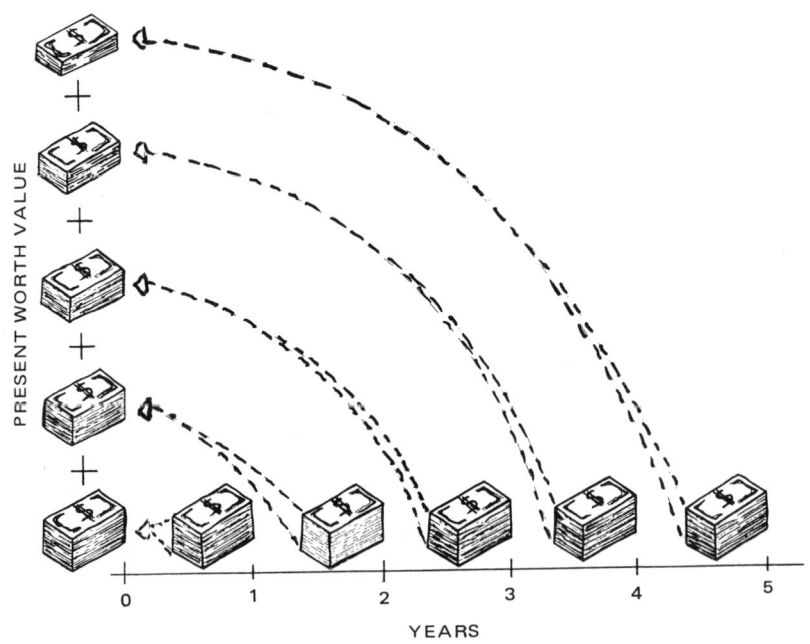

Figure 2-2. Discounting

its initial cost and all subsequent costs, the latter must first be discounted to present value before they are combined with initial cost to obtain the life cycle cost. *It would be erroneous to ignore the timing of the future costs and merely add them to initial cost.*

This last point is extremely important. All too often decision-makers, in order to simplify their work, merely add future costs to present costs. In effect, they are assuming that capital has a time value of zero percent, that is, that capital is free. But capital is not free. Not even a governmental authority which has no debt outstanding can assume free capital since the capital is being taken from the taxpayers and that capital has *their* opportunity rate.

All life cycle cost analysis must be performed in terms of compatible dollars, that is, dollars dated as of a point in time or a period of time. The tools of life cycle cost analysis by which dollar values are shifted in time are six basic interest formulas (explained below). The symbols used in these formulas are:

i = interest rate per period
n = number of interest periods
P = present worth (or present value)
F = future worth (or future value)
A = uniform sum of money in each time period

Cash flow diagrams are included in the illustrations as an aid in the visualization of the cash flows associated with each project. Each cash flow is depicted by an arrow oriented either in an upward or downward direction. An upward directed arrow indicates an income or cash received, while a downward directed arrow indicates an expenditure or outflow of cash. The positions of these arrows or cash amounts are located with respect to time along the horizontal axis while the relative amount of cash expended is indicated by the vertical height of the arrow.

Single Compound Amount (SCA)

$$SCA = (1 + i)^n \qquad \textit{Formula (2-1)}$$

This factor is used to determine the future value of a present investment where interest is compounded at the end of each of n periods at the rate i. SCA represents single compound amount, P is present investment, and F is the value of the investment at the end of the n periods.

$$F = P \times SCA \qquad \textit{Formula (2-2)}$$

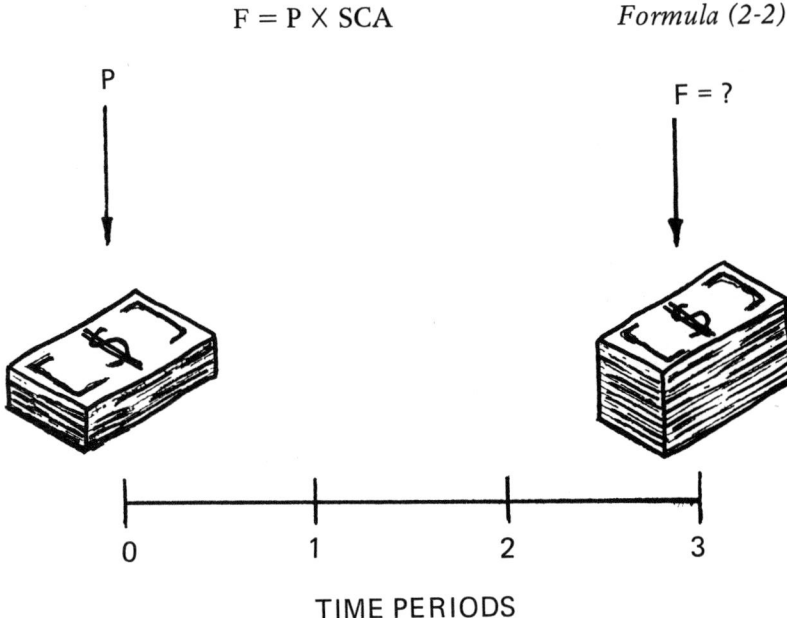

EXAMPLE

Suppose you deposited $1,000 in a savings account which paid interest at the rate of 6% per annum, compounded annually. How much money would you have in the account at the end of 3 years if you made no further deposits or withdrawals? The answer to this question is found by compounding the interest on the principal amount each year for 3 years. The result is called the "future value of a single sum."

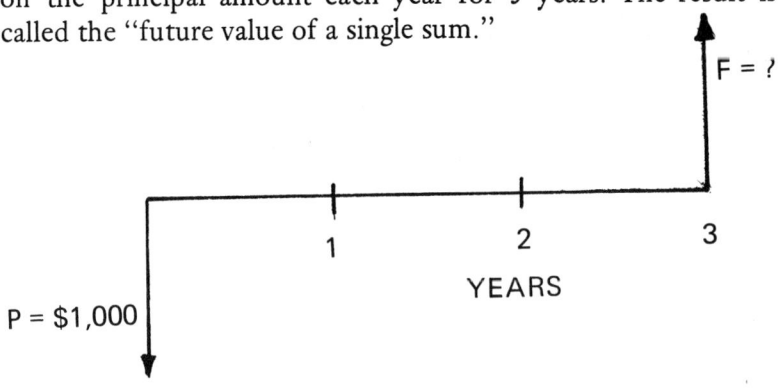

$F = \$1000 \times (1 + .06)^3$
$F = \$1000 \times (1.06)^3$
$F = \$1000 \times 1.19101$
$F = \$1191.01$

Single Present Worth (SPW)

$$SPW = \frac{1}{(1 + i)^n}$$ *Formula (2-3)*

This factor is used to determine the present worth of a future amount discounted at interest rate i for n periods. SPW represents single present worth, P is present investment and F is the future amount to be discounted. Note that SPW is the reciprocal of SCA.

$$P = SPW \times F$$ *Formula (2-4)*

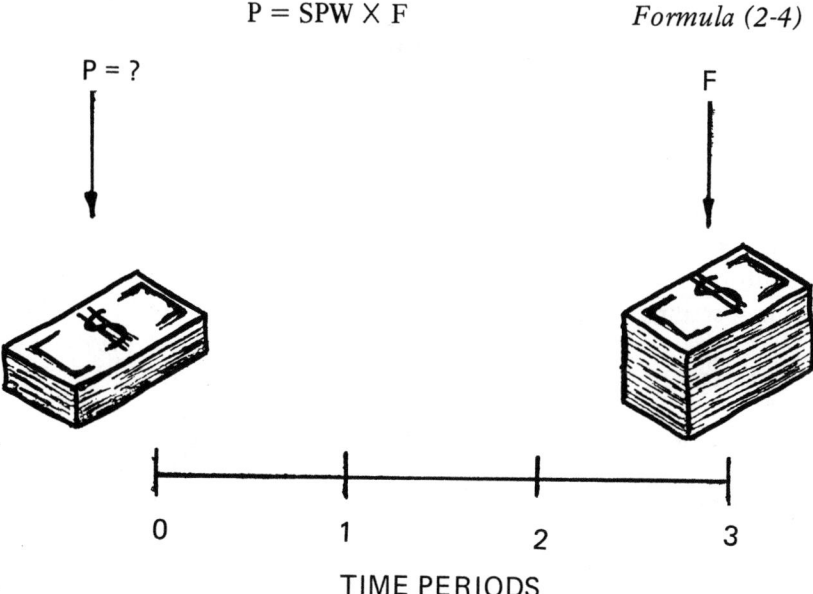

EXAMPLE

How much money would you have to deposit in a savings account today to receive $1,000 at the end of 3 years if the account paid interest of 6% per annum compounded annually? The solution is to find the "present value of a single sum."

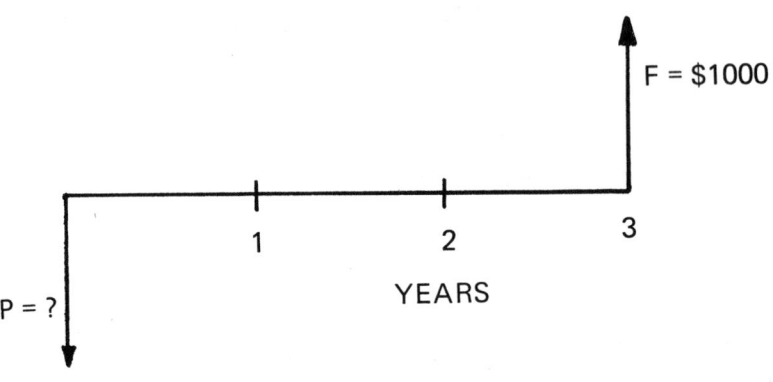

$$P = \$1000 \times \frac{1}{(1 + .06)^3} = \$1000 \times \frac{1}{1.19101}$$
$$P = \$839.62$$

Uniform Capital Recovery (UCR)

$$UCR = \frac{i(1 + i)^n}{(1 + i)^n - 1} \qquad \text{Formula (2-5)}$$

The UCR factor is used where an investment at i% interest rate is returned in n equal periodic installments. UCR represents uniform capital recovery, P represents the investment and A, the installment amount, is the unknown.

$$A = P \times UCR \qquad \text{Formula (2-6)}$$

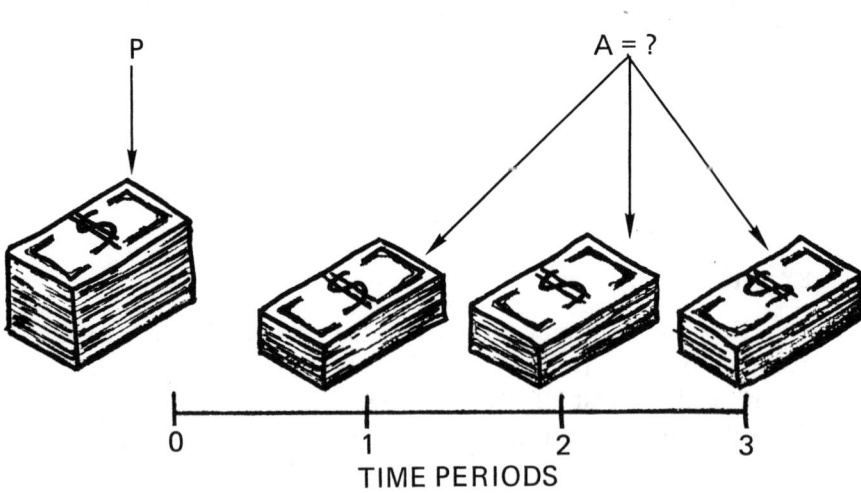

The Time Value of Money

EXAMPLE

Suppose you take out a $20,000, 20-year mortgage on your house, paying 8% interest compounded annually. The uniform capital recovery factor (UCR) will tell you how much your annual mortgage payment would be. In this case, the annual mortgage payment would be $2,037.

$$A = \$20,000 \times \frac{.08(1.08)^{20}}{1.08^{20}-1}$$

$$A = \$2,037.00$$

Uniform Present Worth (UPW)

$$UPW = \frac{(1+i)^n - 1}{i(1+i)^n} \qquad \textit{Formula (2-7)}$$

The UPW factor is used where an investment at i% interest rate is returned in n equal periodic installments. UPW represents uniform present worth, A represents the installment amount, and P, the present worth of the installments, is the unknown. Note that UPW is the reciprocal of UCR.

$$P = UPW \times A \qquad \textit{Formula (2-8)}$$

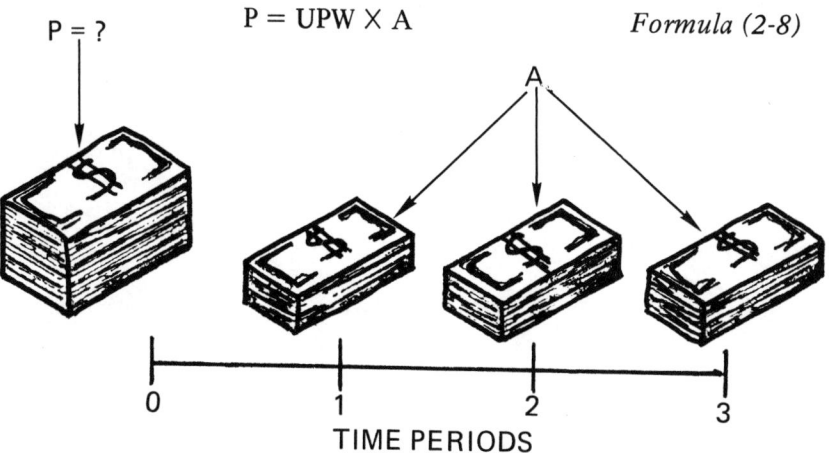

EXAMPLE

What single sum, deposited today at 6% interest compounded annually would enable you to withdraw $1,000 at the end of each of the next 3 years? In other words, we are looking for the "present value of a future annuity."

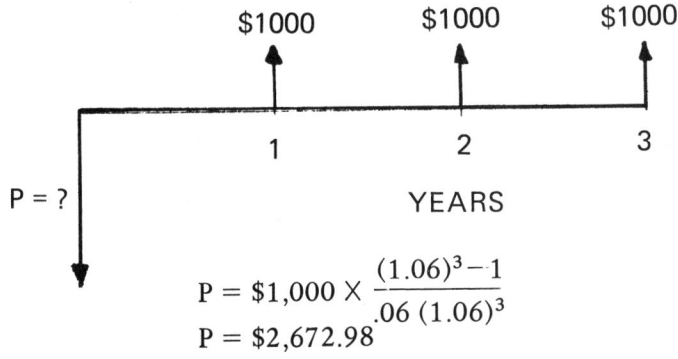

$$P = \$1{,}000 \times \frac{(1.06)^3 - 1}{.06\,(1.06)^3}$$

$$P = \$2{,}672.98$$

Uniform Sinking Fund (USF)

$$USF = \frac{i}{(1+i)^n - 1} \qquad \textit{Formula (2-9)}$$

The USF factor is used where n equal periodic installments, invested at i% interest rate, amount to a known future sum of money. USF represents uniform sinking fund, F represents the known future sum of money, and A, the installment amount, is the unknown. IMPORTANT: The payments are at the *end* of each period.

$$A = F \times USF \qquad \textit{Formula (2-10)}$$

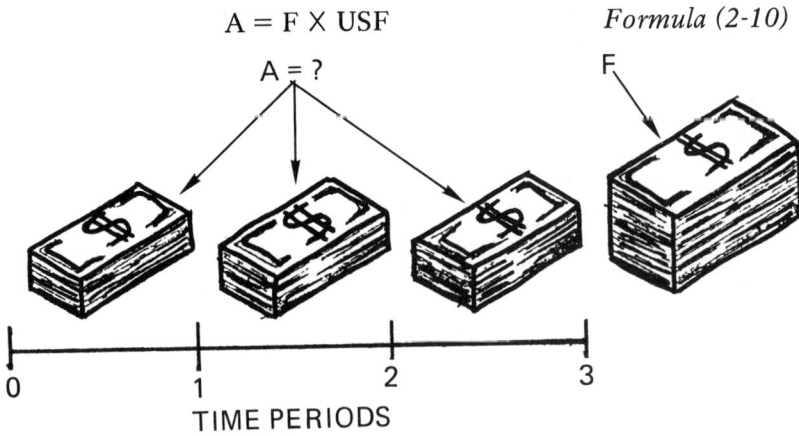

EXAMPLE

The future value of a 3-year annuity earning interest at 6% compounded annually is $100,000. In order to have $100,000 at the end of 3 years, how much do you have to deposit on December 31 of each of the next 3 years in a sinking fund earning 6% interest compounded annually?

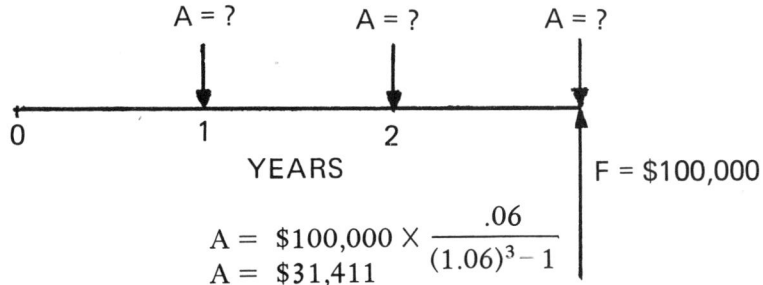

$$A = \$100,000 \times \frac{.06}{(1.06)^3 - 1}$$
$$A = \$31,411$$

Uniform Compound Amount (UCA)

$$UCA = \frac{(1 + i)^n - 1}{i} \qquad \textit{Formula (2-11)}$$

The UCA factor is used where n equal periodic installments, invested at i% interest rate, amount to a future sum of money which is to be determined. UCA represents uniform compound amount, A represents the installment amount, and F, the future sum of money, is the unknown. Note that UCR is the reciprocal of USF.

$$F = A \times UCA \qquad \textit{Formula (2-12)}$$

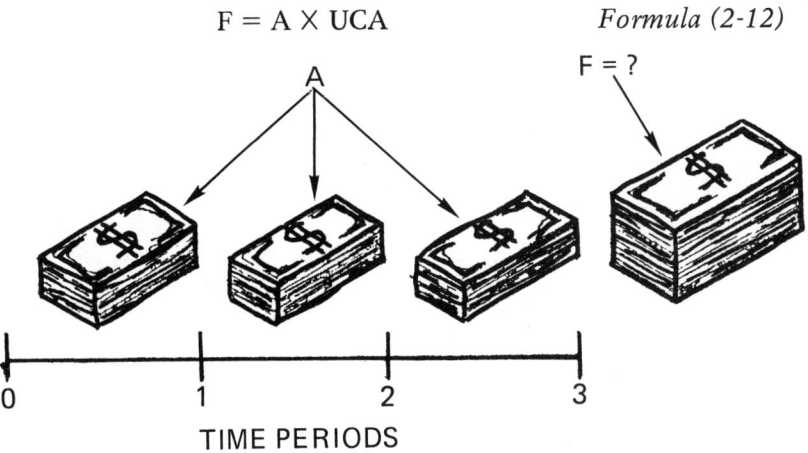

EXAMPLE

If you were to deposit $1,000 on the last day of each of 3 years in a savings account, earning interest at 6% per annum compounded annually, how much would you have at the end of the 3 years, or what would be the "future value of the annuity?" (The word "annuity" is used to describe a series of equal payments made at regular intervals of time.)

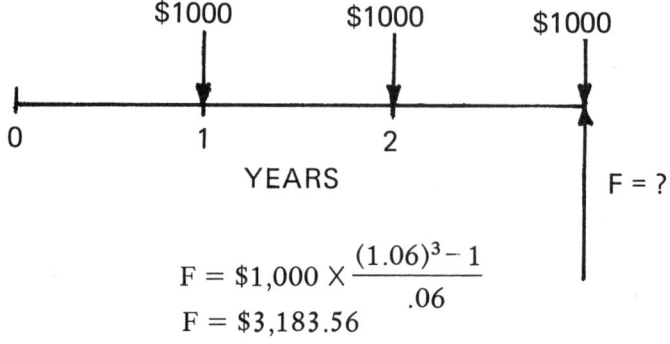

$$F = \$1,000 \times \frac{(1.06)^3 - 1}{.06}$$

$$F = \$3,183.56$$

Use of Interest Tables

To perform the necessary calculations to determine compound interest rates is a lengthy process. As a convenience, Interest Tables using one dollar as the basis of calculation have been included as a separate section in the back of this book.

The following example illustrates how the tables can be used to save time and simplify the compound interest calculation. To find the future worth of $1,000 in 10 years at 10% per year interest rate, use the interest table for Single Compound Amount. The Single Compound Amount Factor is 2.59373 and the future value would be calculated from Formula (2-2):

$$F = P \times SCA$$
$$F = \$1,000 \times 2.59373$$
$$F = \$2,593.73$$

All the other interest values can be obtained in a similar manner from established tables. When an interest rate is not in the table, the formulas may be used or interpolation may be performed.

The Time Value of Money

For example, if the uniform capital recovery factor, UCR, is desired for 7% interest rate at 10 years and only tables for 6% and 8% are available, the interpolation can be performed as follows:

n = 10 years

UCR = @ 6% = .13587
       @ 8% = .14903   UCR @ 7% = .14245

Actual UCR @ 7% from tables is .14233.

FREQUENCY OF COMPOUNDING

The frequency at which the interest rate is compounded has a significant effect on the dollar amount of interest and on the true rate of interest paid. The more frequent the compounding, the greater is the actual interest both in dollars and rate. For example, if an investment pays 8% interest compounded semiannually, the amount and rate are found as follows:

$1 × 8% × ½ year = $ .04
$1.04 × 8% × ½ year = $ .0416
                       .0816

The interest earned on $1 in one year is $.0816 and the interest rate is the equivalent of 8.16% on an annual basis. The 8% rate is referred to as the "nominal rate" and the 8.16% as the "effective rate." A simpler way of converting the 8% nominal rate to the effective rate is to divide the interest rate by 2 and double the compounding period:

$1 × 1.04^2 = $1.0816

Similarly, if the compounding takes place quarterly,

$1 × 1.02^4 = $1.0824

and the effective interest rate is 8.24%.

A comparison of nominal and effective rates for various compounding frequencies (see Figure 2-3):

Nominal Rate	Compounding Frequency	Effective Rate
8.0	Annually	8.0
8.0	Semiannually	8.16
8.0	Quarterly	8.24
8.0	Monthly	8.30
8.0	Continuously	8.33

24 Life Cycle Costing: A Practical Guide for Energy Managers

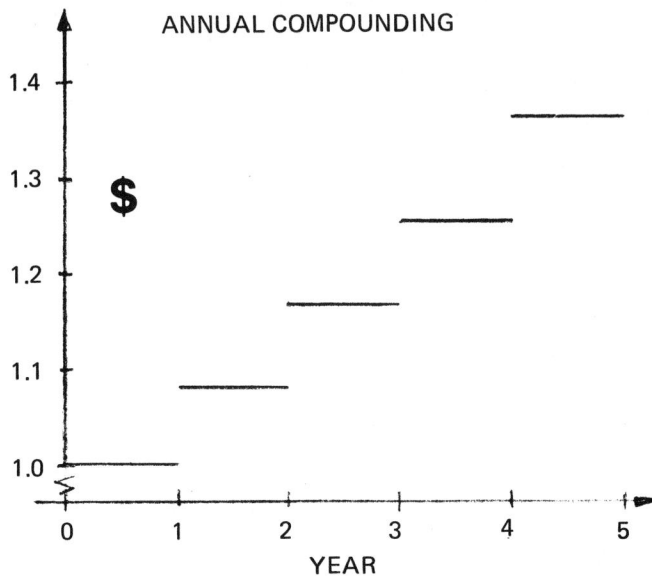

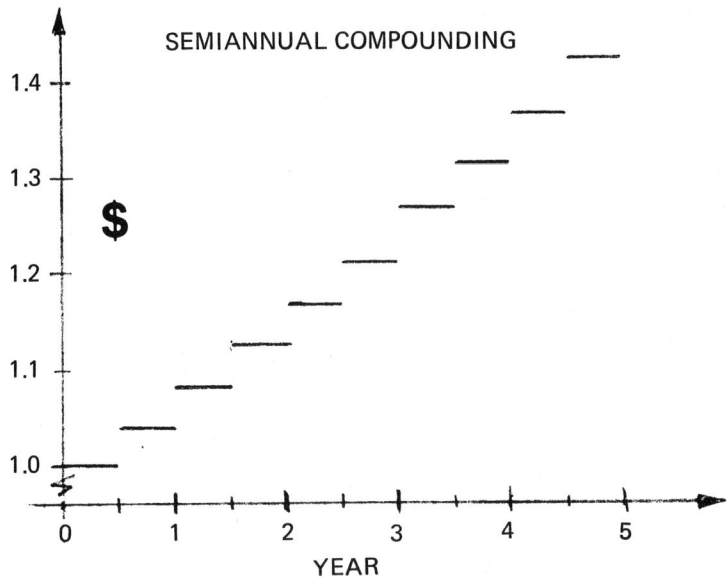

Figure 2-3. $1.00 Compounded at Different Frequencies
(nominal rate = 8%)

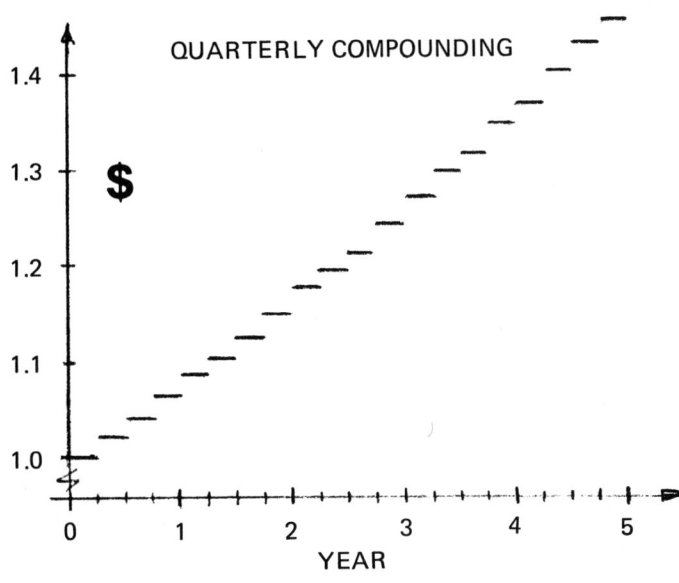

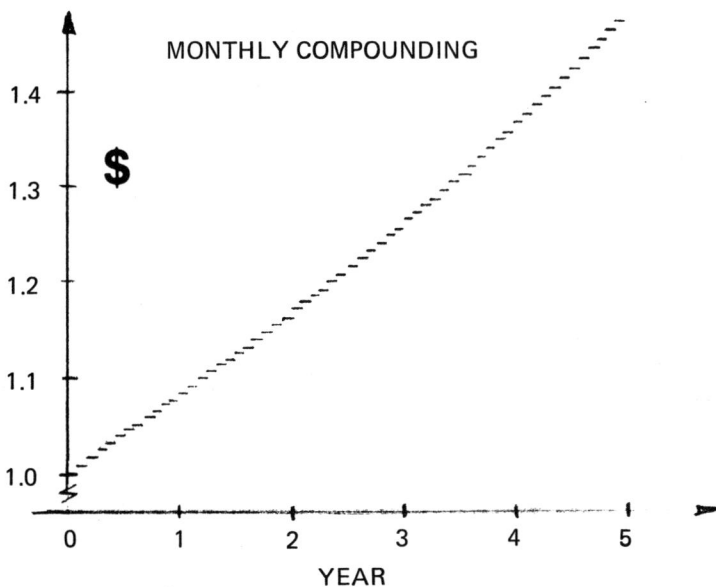

Figure 2-3 (continued)

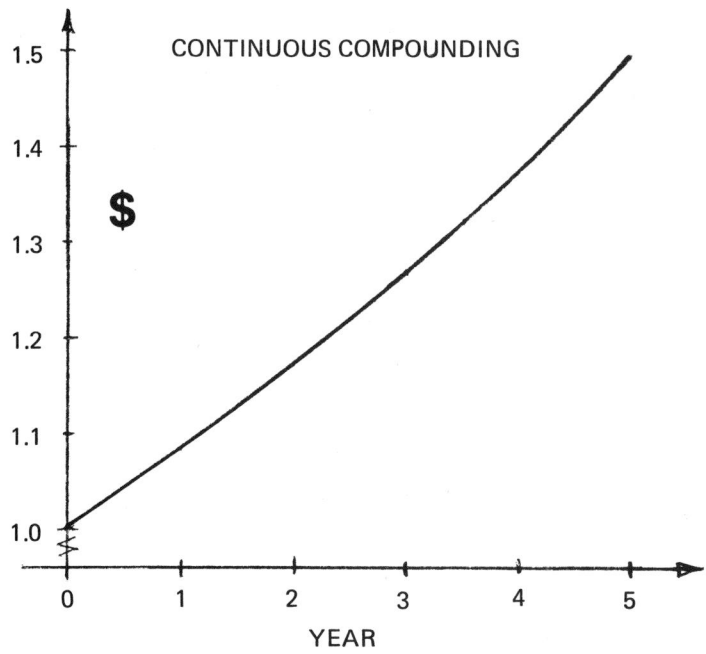

Figure 2-3 (concluded)

The interest rate factor tables provided at the back of this book are based on annual time periods. To use the tables for compounding (and discounting) at more frequent intervals, (1) divide the nominal rate by the frequency of compounding (or discounting) per year and use that interest table, and (2) multiply the frequency of compounding (or discounting) per year by the number of years to find the number of periods.

Example A.—A department store has a carrying charge of 1.25% per month on the unpaid balance in an account. Find the effective rate of interest.

Solution: $(1.0125)^{12} = 1.16$

Answer: $i = 16\%$

Example B.—The quoted interest rate is 10%, compounded semiannually. Find the value of a $1,000 investment at the end of 10 years by using the appropriate interest rate table.

Solution:

F = $1,000 × (SCA, 10 years, 10%) semiannually

The Time Value of Money

$$F = \$1{,}000 \times (SCA, 20 \text{ years}, 5\%)$$
$$F = \$1{,}000 \times 2.65326$$
$$F = \$2{,}653.26$$

CONTINUOUS COMPOUNDING

Compounding of interest can take place at progressively shorter intervals until the limit is reached at continuous compounding. In other words, the frequency of compounding approaches infinity. Using m to represent the frequency of compoundings per year, this concept can be expressed as $m \to \infty$. Letting y represent the number of years, the SCA formula for continuous compounding would become

$$SCA = \lim_{m \to \infty} \left(1 + \frac{i}{m}\right)^{my} \qquad \textit{Formula (2-13)}$$

From calculus:

$$\lim_{x \to \infty} \left(1 + \frac{1}{x}\right)^{x} = e = 2.7182812$$

Let $\dfrac{1}{x} = \dfrac{i}{m}$ and $m = ix$. Substitute in the SCA equation—

$$SCA = \lim_{x \to \infty} \left[\left(1 + \frac{1}{x}\right)^{x}\right]^{iy} = e^{iy}$$

Thus
$$SCA = e^{iy} \qquad \textit{Formula (2-14)}$$
and
$$SPW = \frac{1}{SCA} = e^{-iy} \qquad \textit{Formula (2-15)}$$

CONTINUOUS CASH FLOWS

The SCA and SPW formulas above will accommodate continuous compounding for problems involving single sums of money. Periodic flows of money, however, are a different matter since they can be thought of as occurring in lump sums at specific intervals, that is, discrete payments, or continuously. We present the formulas for both, on the next page, without derivation.

Table 2-1 is useful for determining the values for the continuous compounding interest formulas.

Continuous Compounding Interest Formulas

Formulas	Discrete Flows		Continuous Flows	
UCR	$\dfrac{e^{i(n+1)} - e^{in}}{e^{in} - 1}$	Formula (2-16)	$\dfrac{ie^{in}}{e^{in} - 1}$	Formula (2-17)
UPW	$\dfrac{e^{in} - 1}{e^{i(n+1)} - e^{in}}$	Formula (2-18)	$\dfrac{e^{in} - 1}{ie^{in}}$	Formula (2-19)
USF	$\dfrac{e^{i} - 1}{e^{in} - 1}$	Formula (2-20)	$\dfrac{e^{in} - 1}{i}$	Formula (2-21)
UCA	$\dfrac{e^{in} - 1}{e^{i} - 1}$	Formula (2-22)	$\dfrac{i}{e^{in} - 1}$	Formula (2-23)

The Time Value of Money

Table 2-1. Exponential Values of e^x, where x represents the exponent of e

x	e^x	e^{-x}	x	e^x	e^{-x}
0.00	1.0000	1.000000	0.50	1.6487	0.606531
0.01	1.0101	0.990050	0.51	1.6653	.600496
0.02	1.0202	.980199	0.52	1.6820	.594521
0.03	1.0305	.970446	0.53	1.6989	.588605
0.04	1.0408	.960789	0.54	1.7160	.582748
0.05	1.0513	0.951229	0.55	1.7333	0.576950
0.06	1.0618	.941765	0.56	1.7507	.571209
0.07	1.0725	.932394	0.57	1.7683	.565525
0.08	1.0833	.923116	0.58	1.7860	.559898
0.09	1.0942	.913931	0.59	1.8040	.554327
0.10	1.1052	0.904837	0.60	1.8221	0.548812
0.11	1.1163	.895834	0.61	1.8404	.543351
0.12	1.1275	.886920	0.62	1.8589	.537944
0.13	1.1388	.878095	0.63	1.8776	.532592
0.14	1.1503	.869358	0.64	1.8965	.527292
0.15	1.1618	0.860708	0.65	1.9155	0.522046
0.16	1.1735	.852144	0.66	1.9348	.516851
0.17	1.1853	.843665	0.67	1.9542	.511709
0.18	1.1972	.835270	0.68	1.9739	.506617
0.19	1.2092	.826959	0.69	1.9937	.501576
0.20	1.2214	0.818731	0.70	2.0138	0.496585
0.21	1.2337	.810584	0.71	2.0340	.491644
0.22	1.2461	.802519	0.72	2.0544	.486752
0.23	1.2586	.794534	0.73	2.0751	.481909
0.24	1.2712	.786628	0.74	2.0959	.477114
0.25	1.2840	0.778801	0.75	2.1170	0.472367
0.26	1.2969	.771052	0.76	2.1383	.467666
0.27	1.3100	.763379	0.77	2.1598	.463013
0.28	1.3231	.755784	0.78	2.1815	.458406
0.29	1.3364	.748264	0.79	2.2034	.453845
0.30	1.3499	0.740818	0.80	2.2255	0.449329
0.31	1.3634	.733447	0.81	2.2479	.444858
0.32	1.3771	.726149	0.82	2.2705	.440432
0.33	1.3910	.718924	0.83	2.2933	.436049
0.34	1.4049	.711770	0.84	2.3164	.431711
0.35	1.4191	0.704688	0.85	2.3396	0.427415
0.36	1.4333	.697676	0.86	2.3632	.423162
0.37	1.4477	.690734	0.87	2.3869	.418952
0.38	1.4623	.683861	0.88	2.4109	.414783
0.39	1.4770	.677057	0.89	2.4351	.410656
0.40	1.4918	0.670320	0.90	2.4596	0.406570
0.41	1.5068	.663650	0.91	2.4843	.402524
0.42	1.5220	.657047	0.92	2.5093	.398519
0.43	1.5373	.650509	0.93	2.5345	.394554
0.44	1.5527	.644036	0.94	2.5600	.390628
0.45	1.5683	0.637628	0.95	2.5857	0.386741
0.46	1.5841	.631284	0.96	2.6117	.382893
0.47	1.6000	.625002	0.97	2.6379	.379083
0.48	1.6161	.618783	0.98	2.6645	.375311
0.49	1.6323	.612626	0.99	2.6912	.371577
0.50	1.6487	0.606531	1.00	2.7183	0.367879

Example A.—Compute the present worth of a series of $100 flows to be received at the end of each of the next 5 years with 20% nominal interest compounded continually.

Solution.—The formula to be used is the UPW for discrete flows. Answer:

$$\text{UPW} = \$100 \times \frac{e^{in} - 1}{e^{i(n+1)} - e^{in}} = \$100 \times 2.8552 = \$285.52$$

Example B.—Same as A except that the $100 will be considered as flowing continuously through the years.

Solution.—Use the UPW for continuous flows. Answer:

$$\text{UPW} = \$100 \times \frac{e^{in} - 1}{ie^{in}} = \$100 \times 3.1606 = \$316.06$$

In the interest of accuracy some professionals prefer to work with continuous compounding formulas rather than the discrete formulas. Financial calculators may be used for such computations. In this book the only interest tables provided are for discrete compounding of discrete cash flows. The exponential functions table (Table 2-1) is convenient in using the continuous compounding formulas. It gives values of e with exponents from 0.00 up to 1.00. For simplicity in the presentation of concepts, the examples in the text are limited to problems involving only discrete flows and discrete compounding.

A weakness of this approach is that all costs and benefits are assumed to be occurring at the end of each time period. To illustrate the difference that this can make, let us calculate the present worth of a series of $100 annual costs over 5 years at a nominal interest rate of 20%.

(a) First, assume the costs occur at the end of each year and use the discrete tables.

P = $100 × (UPW, 5 years, 20%) = $299.06

(b) Next, assume the costs occur continuously throughout the 5 years and that the interest is compounded continuously.

P = $316.06 (same calculations as in Example B above).

A way of improving accuracy and still using the discrete tables is to assume that the payments are made at an earlier point in the year rather than at the end. After all, even the middle of the year would be better intuitively since possibly half the costs occurred in each half of the year.

Let us refine the above illustration for solution (a) by assuming that the annual costs will occur two-thirds of the way through each year. Thus:

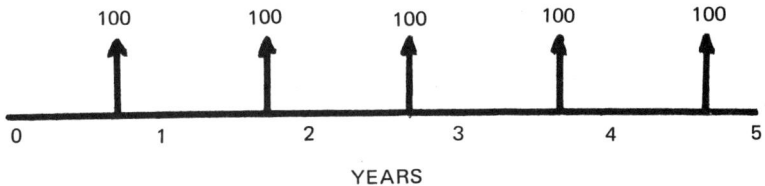

Solve this problem by (a) computing the present worth of payments 2 through 5 treating them as a 4-year annuity, (b) adding to this value the first payment of $100, and (c) calculating the present worth of this sum.

Solution.—
(a) $100 × (UPW, 4, 20%) = $258.87
(b) $258.87 + 100 = $358.87

(c) $358.87 × $\dfrac{1}{1 + (.20 \times 2/3)}$ = $316.65

This result is quite close to the $316.06 found by the continuous compounding method.

Warning.—Continuous compounding tables are available but in such a wide variety of forms that the user must be very cautious using them. There are continuous compounding tables for discrete cash flows and continuous cash flows using both nominal and effective interest rates. That makes four kinds of tables. And there are others. The user should be sure to ascertain which kind of table he or she is using.

There are no continuous compounding tables in this book.

Example C.—If $10,000 is invested today at 10% interest per year, what would it be worth 10 years in the future?

Solution.—Use column 1 of interest tables in the back of the book.

F = P × (SCA; 10, 10%)
F = $10,000 × 2.59373
F = $25,937.30

Example D. — If $3,000 is invested at 14%, how much will be accumulated at the end of 20 years?
Solution. —
F = P × (SCA; 20, 14%)
F = $3,000 × 13.74334
F = $41,230.02

Example E. — What is the present worth of $3,000 to be paid 10 years from today if interest is 9% per year?
Solution. — Use column 2 of the interest tables.
P = F × (SPW; 10, 9%)
P = $3,000 × .42241
P = $1,267.23

Example F. — What is the present worth of $6,000 to be paid 5 years from today if interest is compounded semiannually and is 10% per annum?
Solution. —
P = F × (SPW; 10, 5%)
P = $6,000 × .61391
P = $3,683

Example G. — If $2,000 is borrowed today at 9% interest and it is to be repaid in 5 years with equal annual payments, what could these payments be?
Solution. — Use column 3 of the interest tables.
A = P × (UCR; 5, 9%)
A = $2,000 × .25709
A = $514.18

Example H. — What annual interest rate is being paid if $10,000 is being repaid in equal annual payments of $1,490.30 during a 10-year period?
Solution. —

$$(\text{UCR}; 10, i) = \frac{A}{P}$$

$$(\text{UCR}; 10, i) = \frac{1490.30}{10000}$$

$(\text{UCR}; 10, i) = .149030$

Checking the interest tables, we see that i = 8.00%.

Example I.—What is the present worth of 12 annual end-of-year payments of $350.00 if the interest rate is 13% per year?
Solution.—Use column 4 of the interest tables.
P = A × (UPW; 12, 13%)
P = $350 × 5.91763
P = $2,071.17

Example J.—What is the present worth of $100.00 semi-annual payments for 3 years at 10%?
Solution.—
P = A × (UPW; 6, 5%)
P = $100 × 5.07569
P = $508

Example K.—John knows he needs $5,000 5 years from today. What Uniform Annual Payment must be made to provide it if interest is 9% per year?
Solution.—Use column 5 of the interest tables.
A = F × (USF; 5, 9%)
A = $5,000 × .16709
A = $835.45

Example L.—Tom knows he will need $3,000 in 5 years. He can afford to pay $500 in Uniform Annual Payments. What interest rate must be used?
Solution.—
A = F × (USF; 5, i)
500 = 3,000 × (USF; 5, i)

$$(USF; 5, i) = \frac{500}{3000} = .16666$$

Checking interest tables i ≅ 9%

Example M.—What would 10 uniform end-of-year payments of $300.00 per year be worth in 10 years if the interest rate is 11% per annum?
Solution.—Use column 6 of the interest tables.
F = A × (UCA; 10, 11%)
F = $300 × 16.72191
F = $5016.57

Example N.—Mary needs $1,000 on her twenty-first birthday. Today is her eighteenth birthday. She can afford uniform

payments of $300.00 per year at the end of each year. What interest rate must Mary invest her $300.00 annual payments?
Solution.—
F = A × (UCA; 3, i)
1,000 = 300 × (UCA; 3, i)

$$(UCA; 3, i) = \frac{1000}{300} = 3.33$$

Checking interest tables i ≅ 11%

SELF-STUDY PROBLEMS

Throughout the text problems are presented to help you master the principles discussed. Solutions are given in a special section at the back of the book.

Problem 2-1.—$1,000 is paid at the end of each of the next 5 years. What is the present worth of these payments if interest is computed at 8% per year?

Problem 2-2.—A lender is willing to advance $30,000 on a 12% 20-year mortgage with annual payments made at the end of each year. What is the annual payment? How would this problem be solved if the payments were made monthly?

Problem 2-3.—Jim Evans has the following debts:
    a.   20 annual mortgage payments of $2,400,
    b.   12 bimonthly payments of $75 on his automobile,
    c. $6,000 in debts due in 2 years,
    d. $700 due today.

Using an annual interest rate of 12%, calculate the uniform annual amount necessary to retire all of the debts if Jim refinances the debts for 20 years.

Problem 2-4.—Carol borrowed $5,000 from her uncle. The loan is to be repaid at the end of 4 years at an interest rate of 14% per year, compounded annually.
    a.   How much will Carol owe in 4 years?
    b.   How much will Carol owe if the interest rate is compounded semiannually?

Problem 2-5.—Bob has $12,000 in cash today. He knows that in 4 years he will need $8,000 for a new car. Bob wants to take a vacation to Bermuda with his family. Bob's savings account pays interest at 8% per annum, compounded quarterly. How much can Bob spend on his vacation and still be able to buy his new car in 4 years?

Problem 2-6.—Mr. Smith earns $15,000 per year after taxes. Mrs. Smith earns $10,000 per year after taxes. The Smith living expenses amount to $15,000 per year and the remaining earnings are placed in a savings account. How much will the Smiths have in their savings account at the end of 5 years? Deposits are made in equal amounts on a quarterly basis and interest is compounded quarterly at 8% per annum.

Problem 2-7.—If John needs $5,000 5 years from today, what is the uniform annual payment John must make if interest is 10% per year?

Problem 2-8.—How much would Mary have to pay per year if she wants to have $3,000 9 years from today and her first payment is to be today? Interest is 5% per year.

Problem 2-9.—Mr. John Smith wishes to provide his alma mater with enough funds to grant a $5,000 scholarship award every year for 20 years. How much money must he donate now in order to accomplish this if the university can invest this money at 5% per annum? How many such scholarship awards can he endow by donating $250,000 to the university today? Assume the tuition will remain the same for the next 20 years.

Problem 2-10.—What is the present value of a promise to pay $100 at the end of each of the next 10 years if interest is computed at 8%?

Problem 2-11.—How long will it take money to double at 5%? At 6%? How long will it take money to triple at the above rates?

Problem 2-12.—At what interest rate must a man invest $159 per year in order to have a balance of $2,000 after the 10th payment?

Problem 2-13.—What is the value today of a $10,000 gift to be made 15 years from now, if the time value of money is 10%?

Problem 2-14.—What interest will be paid on a $3,000 loan, assuming an interest rate of 5% compounded annually and the loan paid in one lump sum at the end of a 4-year period?

Problem 2-15.—Hector borrowed $1,000 from his friend George, promising to repay the loan in 5 years, plus interest at 6% per annum, compounded annually.

    a.   How much will George receive in 5 years?
    b.   How much would George receive if the original loan had been for 3 years?
    c.   How much will George receive if the interest rate were 7%?
    d.   Answer question 2c assuming that interest is compounded semiannually at a rate of 6%.

3

Application of Life Cycle Costing

An oversimplified illustration of a life cycle analysis was presented in Example 1-1 of chapter 1. In this chapter the same problem (the purchase of a car) will be solved properly by allowing for the time value of money.

Example 3-1
(Based on Example 1-1, chapter 1)

Assume the cost of borrowing the money (the discount rate) is 8%.

Let us first apply the *Present Worth Method* to the solution.

	CAR A	CAR B	CAR C
Purchase price	$3,200	$3,800	$3,910
Sales tax	160	190	195
Initial cost	$3,360	$3,990	$4,105
Present value of Annual cost	CAR A	CAR B	CAR C
Fuel	$2,428	$1,822	$1,560
Maintenance	662	464	331
License, Insurance	712	811	1,163
PV of total annual costs	$3,802	$3,097	$3,054
Initial cost	$3,360	$3,990	$4,105
	$7,162	$7,087	$7,159
Deduct Trade-in	(74)	(96)	(147)
Total LCC (P.V.) =	$7,088	$6,991	$7,012
Differential	$ 97	—	$ 21

Best Choice: B, if only dollar costs are considered; A or C if the anticipated benefits are worth the differential.

SUPPORTING CALCULATIONS

Fuel
A. Present value of $733 a year for 4 years discounted at 8% = $733 × 3.31212 = $2,428
B. $550 × 3.31212 = $1,822
C. $471 × 3.31212 = $1,560

Maintenance
Assume the tune-ups are paid at the end of each year and that we are discounting for the average yearly tune-up cost as found previously.
A. $200 × 3.31212 = $662
B. $140 × 3.31212 = $464
C. $100 × 3.31212 = $331

Licenses and Insurance
A. $215 × 3.31212 = $712
B. $245 × 3.31212 = $811
C. $351 × 3.31212 = $1,163

Salvage Value (Trade-in)
Present value of a single amount, discounted for 4 years.
A. $100 × .73503 = $ 74
B. $130 × .73503 = $ 96
C. $200 × .73503 = $147

Conclusion.—LCC may determine the present values of the alternatives and the dollar differentials. These figures, however, do not provide decision-makers with average *annual* dollar differentials. The present value differential between Cars B and C is $21. If the decision-maker were interested in determining the average annual dollar differential, he would proceed as follows:

a. Convert original outlays to an average annual outlay by multiplying by the capital recovery factor.
b. Add average annual outlays for costs to the above.
c. Convert irregular outlays to present value and then treat as item a.
d. Convert trade-in value to an average annual benefit by multiplying by the sinking fund factor and then subtracting from the totals of a, b and c.

Application of Life Cycle Costing

Now let us consider the car problem using the *Uniform Annual Cost Method*.

	CAR A	CAR B	CAR C
Purchase Price	$3,200	$3,800	$3,910
Sales Tax (5%)	160	190	195
Initial Cost	$3,360	$3,990	$4,105
(UCR 4, 8%) ×	.30192	.30192	.30192
Average annuals	$1,014	$1,205	$1,239
Fuel	$ 733	$ 550	$ 471
Maintenance	200	140	100
License and Insurance	215	245	351
Annual Costs	$1,148	$ 935	$ 922
Salvage (Trade-in)	(100)	(130)	(200)
(USF 4, 8%) ×	.22192	.22192	.22192
	(22)	(29)	(44)
Average Annual LCCs	$2,140	$2,111	$2,117
Average Annual Differentials	29	–	$ 6

Alternative method of calculating average annual cost: Multiply present value LCCs by UCR (Uniform Capital Recovery).

	$7,088	$6,991	$7,012
	.30192	.30192	.30192
Average Annual LCCs	$2,140	$2,111	$2,117

Both approaches are discussed in detail in the following sections.

PRESENT WORTH METHOD

When the "present worth method" of life cycle costing is used, all expenditures, regardless of when they occur, are compared during a common year; that is, baseline year. Future expenditures are properly discounted to reflect their time value. The six basic formulas discussed in chapter 2 must be used to properly discount these future expenditures to their present worth. Once these future expenditures are discounted, they may be compared properly to expenditures incurring "today," or during the "baseline year." Once this discounting is accomplished, all expenditures are weighed on a common basis and may be added together to obtain a total present worth value.

Example 3-2

A HVAC system is expected to cost $50,000. A one-time replacement is expected after 15 years at a cost of $20,000. Annual operating costs are to be $5,000 per year. The system is expected to have a salvage value of $10,000 after 30 years. Using a 10% discount rate, what is the Total Present Worth of the system over 30 years?

SOLUTION

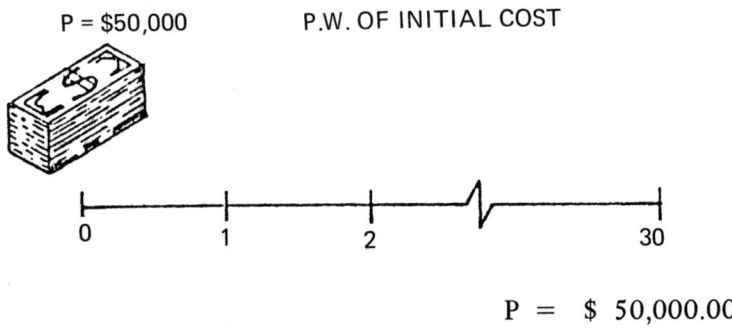

P = $ 50,000.00

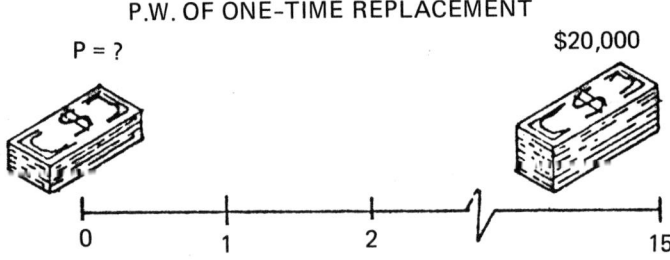

P = $20,000 × (SPW; 15, 10%)
 = 20,000 × .23939

P =   4,787.80

Application of Life Cycle Costing 41

P.W. OF OPERATING COSTS

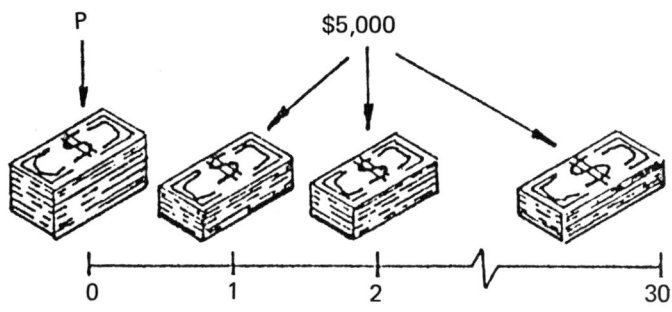

P = $5,000 × (UPW; 30, 10%)
 = 5,000 × 9.42691

P = $ 47,134.55

P.W. OF SALVAGE VALUE

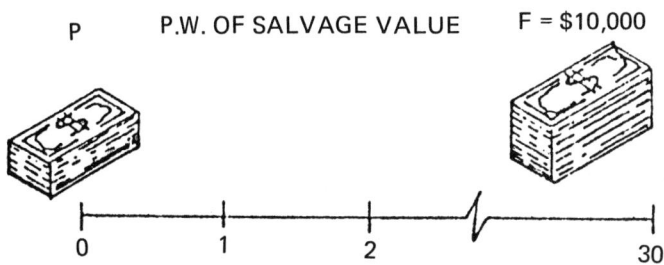

P = $10,000 × (SPW; 30, 10%)
 = 10,000 × .05731

P = (573.10)

Total P.W. of System = $101,349.25

Example 3-3

Alternative A has an initial cost of $20,000, an estimated life of 20 years, an annual operating cost of $2,000 and a salvage value of $2,000.

Alternative B has an initial cost of $21,000, an estimated life of 20 years, annual operating cost of $1,800 and a salvage value of $2,500. The appropriate interest rate for comparing these alternatives is 9% per year.

SOLUTION

Alternative A.—
P.W. initial cost			= $20,000
P.W. operating costs	= 2,000 ×	(UPW; 20, 9%)	
	= 2,000 ×	9.12852	= 18,257
P.W. salvage	= 2,000 ×	(SPW; 20, 9%)	
	= 2,000 ×	.17843	= (357)
Total P.W.			$37,900

Alternative B.—
P.W. initial cost			= $21,000
P.W. operating costs	= 1,800 ×	(UPW; 20, 9%)	
	= 1,800 ×	9.12852	= 16,431
P.W. salvage	= 2,500 ×	(SPW; 20, 9%)	
	= 2,500 ×	.17843	= (446)
Total P.W.			$36,985

Differential in favor of B = $915.00

UNIFORM ANNUAL COST METHOD

The "uniform annual cost method" of calculating life cycle costs reduces each alternative cost to the equivalent base of a uniform annual cost. By using this method, both present dollars and future dollars are converted to a uniform annual cost while taking into account the time value of money at a particular interest rate. The six basic formulas discussed in chapter 2 must be used to properly establish these annual costs.

All present costs are broken down into equivalent yearly payments throughout the life cycle. All future costs spent at any time during the life cycle, are also broken down into equivalent yearly payments throughout the life cycle. All the equivalent yearly costs are then added together to establish the total uniform annual cost.

When comparing alternatives, the same choice will be made regardless of whether the present worth method or uniform annual cost method is used. The same relative cost advantage will result from either method of calculation.

Examples 3-4 and 3-5

Solve Examples 3-2 and 3-3 using the Uniform Annual Cost Method.

SOLUTION

ANNUAL OWNING COSTS

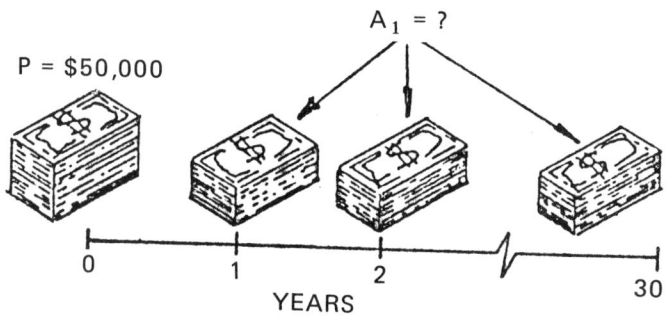

$$A_1 = \$50,000 \times (UCR, 30, 10\%)$$
$$= 50,000 \times .10608$$
$$= \qquad \qquad \$5,304.00$$

ANNUAL FUEL COSTS

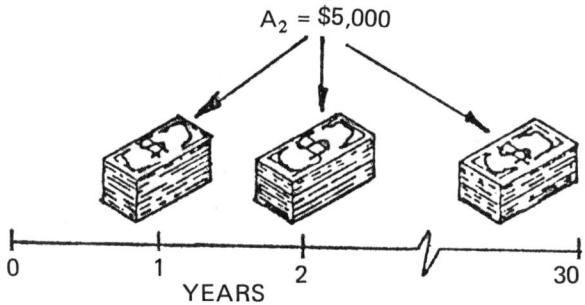

$$A_2 = \text{Fuel Cost}$$
$$A_2 = \$5,000 \qquad \qquad \$5,000.00$$

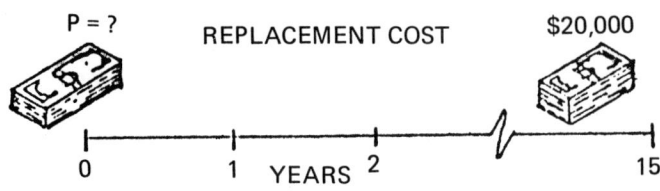

REPLACEMENT COST

$$P = F \times SPW = \$20{,}000 \times .2394 = \$4{,}788$$

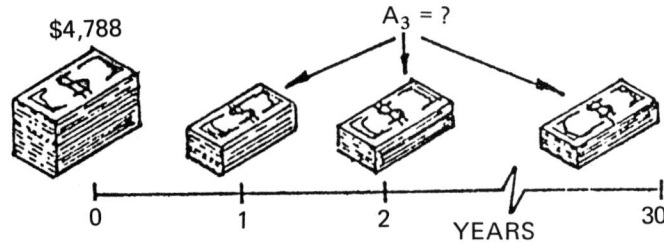

$$\begin{aligned} A_3 &= \$4{,}788 \times (UCR, 30, 10\%) \\ &= 4{,}788 \times .10608 \\ &= \$\ 507.91 \end{aligned}$$

ANNUAL SALVAGE VALUE

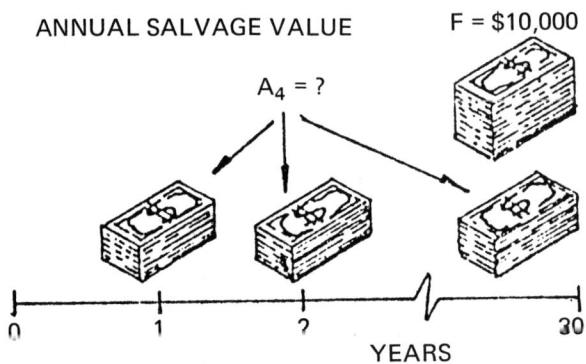

$$\begin{aligned} A_4 &= \$10{,}000 \times (USF, 30, 10\%) \\ &= 10{,}000 \times .00608 \\ &= (\$60.80) \end{aligned}$$

Annualized Cost = $10,751.11

Example 3-6

Alternative A has an initial cost of $20,000, an estimated life of 20 years, an annual operating cost of $2,000, and a salvage value of $2,000.

Alternative B has an initial cost of $21,000, an estimated life of 20 years, annual operating cost of $1,800, and a salvage value of $2,500. The appropriate interest rate for comparing these alternatives is 9% per year.

SOLUTION

Alternative A.—

A_1 initial cost	= P × (UCR; 20, 9%)	
	= $20,000 × .10955	= $2,191
A_2 annual operating costs		= 2,000
A_3 salvage value	= F × (USF; 20, 9%)	=
	= $2,000 × .01955	= (39)
	Uniform Annual Cost	= $4,152

Alternative B.—

A_1 initial cost	= P × (UCR; 20, 9%)	
	= $21,000 × .10955	= $2,300
A_2 annual operating costs		= 1,800
A_3 salvage value	= F × (USF; 20, 9%)	
	= $2,500 × .01955	= (49)
	Uniform Annual Cost	= $4,051

Conclusion: Alternative B is a better choice.

FOREVER PROJECTS

A "forever project" is one which requires outlays over a lengthy, indefinite planning horizon. An example would be the acquisition of a piece of land for a public park. Maintenance costs would be expected to continue indefinitely.

To determine the present worth of equal annual costs of a forever project, the costs are treated as an infinite series. To convert an infinite series of equal annual costs to present worth, divide by the discount rate.

Example 3-7

What is the present worth of a series of $1,000 costs forever if the discount rate is 10%?

SOLUTION

$$\$1{,}000 \div .10 = \$10{,}000 \quad \text{Answer}$$

A more familiar way of expressing this problem is: What amount of money, if left invested at 10%, will pay $1,000 a year forever? The answer, of course, is $10,000.

The uniform annual cost method is often more convenient for analyzing forever projects. Initial cost is converted to an average annual cost by multiplying by the discount rate. Costs occurring at regular intervals of several years are converted by multiplying by the USF (uniform sinking fund) factor.

Example 3-8

Two plans for a program that will continue indefinitely are under study.

Plan A has an initial investment of $20,000. In 10 years, and every 10 years thereafter, the renewal costs will be $15,000. Every 3 years, there will be a cost of $2,000. Annual operating cost will be $1,000.

Plan B requires an initial investment of $35,000 and an additional $25,000 every 15 years. Annual operating costs are $700. The interest rate appropriate to this study is 10%.

SOLUTION

Plan A.—

$20,000 × .10	= $2,000
*15,000 × (USF, 10, 10%)	941
2,000 × (USF, 3, 10%)	= 604
$ 1,000	1,000
Uniform Annual Cost	= $4,545

Plan B.—

$35,000 × .10	= $3,500
25,000 × (USF, 15, 10%)	= 787
$ 700	700
Uniform Annual Cost	= $4,987

* Alternate Method:
  $15,000 × (SPW, 10, 10%) = $5,783
  5,783 × (UCR, 10, 10%) = $ 941

SELF-STUDY PROBLEMS

Problem 3-1.—An industrial manufacturer wants to investigate the economics of investing in new machinery for its production plants. The initial cost of this new machinery will be $95,000, with an estimated life of 20 years. If the equipment is purchased, the Vice-President of Manufacturing estimates that operating costs for the first 10 years will be reduced by $12,000 per year. For the next 5 years, annual operating costs will be reduced by $5,000 per year, and the remaining 5 years operating costs will be reduced by $3,000 per year.

The expected salvage value at the end of 20 years is $10,000. The appropriate interest rate is 10%.

A graphic representation of the costs and savings will help to solve this problem:

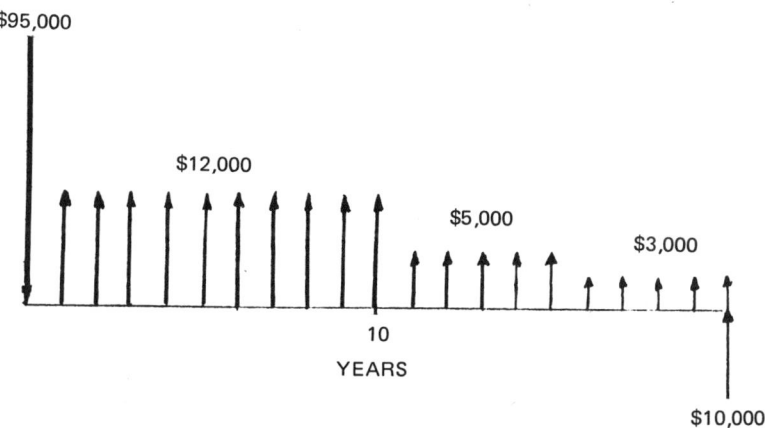

Is it advisable to invest in this project?

Problem 3-2.—The federal government is evaluating the possibility of initiating a mass inoculation program for the immunization of a threatening disease. The disease must be analyzed with economy in mind. The federal government must determine if a $2,000,000 investment with an annual operating cost of $200,000 for this immunization program is worthwhile.

This program is expected to save $500,000 per year to the taxpayer.

With a 12% interest rate, is this program economically feasible?

Problem 3-3.—The Defense Department is interested in establishing the life cycle cost of a low pressure chamber to be used for testing. Initial cost is $240,000. The chamber will be scheduled each year for 3600 hours less downtime. The expected MTBF (mean time between failures) is 300 hours and the MTTR (mean time to repair) is 60 hours. The labor cost for repairs is $10 per hour. One man is needed to operate the chamber and he will have to be paid $8 an hour for the entire 3600 hours, even though he will be idle while the chamber is being repaired. Power required is 10 KW at a cost of $.03 per KWH (only for hours in use). The Department wants the use of the chamber for 3 years after which the unit will be sold at an expected salvage price of $100,000. The discount rate is 10%.

4
Costs

CLASSIFICATION

An important step in life cycle cost analysis is the development of an all-inclusive listing of all costs and dollar benefits associated with a project. This listing must include all dollar flows from the first outlay to disposal of the item and must be categorized by stage of ownership to facilitate the LCC analysis.

Example: Equipment

1. *Initial Cost Period.*—Conceptual studies, design, planning, acquisition, installation, testing, training, financing of preoperational phase. Expenditures made prior to the LCC analysis (called "sunk costs") are not included.
2. *Use Period.*—
    a. Operating costs—staffing, energy, insurance, security, increases in working capital, personnel training.
    b. Maintenance costs—parts used, repair, cleaning, painting, all corrective and preventive maintenance.
    c. Replacement costs.
3. *Disposal Period.*—Costs of removal and restoration; salvage value and release of working capital are benefits (i.e., negative costs).

INTEREST COST

In the problems encountered thus far in this text, annual interest expenditure has not been included. A question may arise in the mind of the reader as to whether this was due to an oversight, a simplification of the problem, or an unstated assumption that the purchaser had sufficient funds and did not have to raise money from another source. The answer is that none of these explanations are correct. The annual interest was accounted for in the discounting procedure.

To illustrate this point, let us calculate the present worth of a car with and without the financing cost included. Initial cost is $5,000, annual costs are $1,000, life is 5 years, and there is no salvage value. The borrowing rate is 10%.

Method A: Loan Not Included

Initial Cost	$5,000.00
P.W. of Annual Costs =	
(UPW; 5, 10%) × 1000 =	3,790.77
Total P.W. of Costs	= $8,790.77

Method B: Loan Included

Initial Cost	$5,000.00
Loan Received	(5,000.00)
P.W. of Annual Costs	= 3,790.77
P.W. of Interest Charges:	
(UPW; 5, 10%) × 500 =	1,895.39
P.W. of Loan Repayment:	
(SPW; 5, 10%) × 5000 =	3,104.60
Total P.W. of Costs	= $8,790.76

The answer is the same by both methods. Method A, however, involves fewer calculations and is the method used in this book.

It is important to understand, moreover, that even if the purchaser does not need to raise funds the appropriate discount rate should be determined and used in the calculations. After all, there is a cost (opportunity cost) to using one's own funds since the return which would be earned by investing the funds elsewhere is being forfeited. This applies even in the public sector where the funds may be available from tax revenues.

DEPRECIATION

Another question which may arise in the mind of the reader is: Shouldn't the wear (depreciation) on a piece of equipment be included as a cost?

The answer is no, since there is no cash outlay for depreciation. The cash was paid initially for the purchase and to include depreciation as a cost would be counting it twice. Purchasers in the private sector who are subject to federal income tax may deduct depreciation as an expense, provided an approved method of accounting for the depreciation is used.

FORMAT

A typical, although simplified, format for life cycle cost analysis is provided on pages 52 and 53. The user may find the form useful as shown, or may wish to modify it for specific applications.

COST DETERMINATION

One of the objections to the use of life cycle costing which is often heard is that the estimation of future costs is too difficult. Aren't the estimates merely guesses? And why should decisions be based on guesses? If life cycle costing is based on dubious estimates, how useful can it be?

In response to these questions, it can be argued that there is no alternative to life cycle costing. Buying on initial cost is not an alternative to buying on life cycle cost. All purchases are made for a life cycle and the only difference between initial cost and life cycle cost purchasing is in the judgment about treatment of future costs. In the case of initial cost buying, the judgment is made that a zero value should be attributed to post-purchase costs. Life cycle costing, if done correctly, recognizes that post-purchase costs matter, that they do have a value, and that a considered estimate of those costs is likely to be closer to true value than the zero value presumption. In either case, a judgment about future costs cannot be avoided. The only real issue is what the judgment should be.

Undeniably there is some guesswork in predicting costs, but often information can be obtained which will reduce the uncertainty of the estimates. It may be that an estimate can only be narrowed down to a range, but even that information is useful. If a cost range can be estimated at $300 to $400, for example, a life cycle cost analysis can be performed at both

LIFE CYCLE COST ANALYSIS — SHORT FORM
Using Present Worth and Uniform Annual Cost Methods

Project Name: _____ Life Cycle: _____ yrs.
Location: _____ Interest Rate: _____ %
Owner/Engineer: _____

Escalation Rate: ____ % ____

		Alternate No. 1	Alternate No. 2	Alternate No. 3
Initial Costs	1. Initial Costs			
	a. Base Cost			
	b. Interface & Auxiliary Costs			
	I. _____			
	II. _____			
	III. _____			
	IV. _____			
	V. _____			
	c. TOTAL INITIAL COST			
	d. Difference in Initial Cost			
Total Life Cycle Costs (Present Worth)	2. Operating Cost, P.W.			
	a. Fuel Cost			
	b. Operating Labor Cost			
	c. Maintenance Cost			
	d. Replacement Cost			
	e. Salvage Value ( )			
	f. _____			
	g. _____			
	h. _____			
	i. TOTAL OPERATING COST			
	3. TOTAL P.W. OPERATING & INITIAL COST			
Uniform Annual Costs	4. Uniform Initial Cost			
	5. Uniform Operating Cost			
	6. Total			

LIFE CYCLE COST ANALYSIS
Supplement

The following is a step-by-step description for completing the Life Cycle Cost Analysis — Short Form.

1. **Initial Costs**
   a. Base Cost — Enter basic initial cost of equipment.
   b. Interface & Auxiliary Costs — Enter cost of such items as:
      I. Additional Construction
      II. Additional Plumbing
      III. Additional Electrical
      IV. Fuel Storage Tanks
      V. Any other costs associated with the item under construction
   c. Total Initial Cost = Total of 1a and 1b
   d. Difference in Initial Cost — Subtract low initial cost from high initial cost alternative from respective alternatives

2. **Operating Cost, P.W.**
   a. Fuel Cost = Present Fuel Cost × Discount–Escalation Factor.
   b. Operating Labor Cost = Present Labor Cost × Discount–Escalation Factor.
   c. Maintenance Cost = Present Maintenance Cost × Discount–Escalation Factor.
   d. Replacement Cost = Future Replacement Cost × S.P.W.
   e. Salvage Value = Future Salvage × S.P.W.
   f. Add whatever annual costs that are not considered
   g. above. Tax benefits from depreciation and from ex-
   h. pense deductions may be included here but must be *deducted* from costs.
   i. TOTAL OPERATING COST = 2a + 2b + 2c + 2d − 2e + 2f + 2g + 2h

3. Total Operating & Initial Cost = 1c + 2i
4. Uniform Initial Cost = 1c × UCR
5. Uniform Operating Cost = 2i × UCR
6. Total = 4 + 5

ends of the range and at intermediate levels. The results will aid in the decision process.

In some circumstances cost estimation may not be needed. If a decision is to be made between competing projects and a particular cost is common to those projects it may be omitted from the calculations. The decision will be unchanged. For example, let's say Projects A and B will have identical labor costs:

	Project A	*Project B*
Initial Cost	$ 7,000	$ 8,000
P.W. of energy costs	6,200	5,000
P.W. of labor costs	8,000	8,000
Total life cycle costs	$21,200	$21,000
Differential		200

If the labor cost is omitted from the calculations the differential will still be $200.

INFORMATION SOURCES

An important start toward cost estimation is the collection of pertinent information. A partial listing of sources is presented here.

1. Data bases self-generated or shared.
2. Manufacturers' test results and results of impartial testing services.
3. Trade associations, trade publications. For example:
   Edison Electric Institute (electric rates)
   American Gas Association (gas prices)
   American Petroleum Institute (oil prices)
   National Coal Association (coal prices)
   American Metal Market (metal prices)
   Engineering News-Record (construction costs)
4. Handbooks. For example:
   Building Owners and Management Association (operational and maintenance costs)
5. Indexes. For example:
   Boeckhs, Richardson, Means, Dodge, Building Cost File
6. Government. For example:
   Department of Energy

Department of Labor
Department of Commerce
7. Other. For example:

The American Society of Heating, Refrigeration and Air Conditioning Engineers' sponsored Research Project 186 "Equipment Life and Maintenance Cost Survey" (service life data)

Note: A comprehensive categorization of building costs has been developed by the U.S. General Services Administration and the American Institute of Architects. This model, known as UNIFORMAT, is outlined in Appendix B.

COST ESTIMATING

The subject of cost estimation is treated at length in various works on the subject but the emphasis is often on design or construction. A few of the basic notions of cost estimation which could be of help to purchasers are explained in this section.

A cost estimating relationship (generally referred to as a CER) is any equation that facilitates estimation of a cost.. CERs are of two types: those which are built up from inputs such as material costs, wage rates, time and motion studies, etc., and those which relate to output characteristics such as speed, size, power, capacity, etc. The process using the latter type of CER is referred to as parametric costing and it is used heavily by the Department of Defense for verification of input cost aggregates and for quick estimation of costs where change of equipment design is contemplated.

The "input" type CER is constructed from the elements which compose the particular cost. A simple illustration of a linear estimate:

$$C_L = W_1 \times H_1 + W_2 \times H_2 + F$$

where C_L = cost of labor per year

W_1 and W_2 = hourly wage rates for employees of types 1 and 2

H_1 and H_2 = hours of employment per year for employees of types 1 and 2

F = salary of managerial employee

In this equation the labor cost is composed of wages based on hours worked at a specific rate plus a fixed cost for a managerial employee.

In the above CER, W and H may be constants, variables, or parameters, depending upon the particular case. If W and H are fixed for the year, they are constants. If they are not fixed, they are generally classified as parameters if they can be set arbitrarily by management, or variables if they are controlled by other forces. In the equation, if W and H are variables they are referred to as independent variables and C_L is considered the dependent variable.

Cost estimation is largely a subjective procedure, but it may be supported or facilitated by experience. The mathematical methods used vary from simple arithmetic to complex statistical procedures. Illustrations follow.

Expert Judgment Method

The wage rate for repair of equipment is $10 an hour. We will estimate the annual labor cost of one machine given the information below. Model: Annual labor cost = Number of repairs in a year × length of time per repair × wage rate.

1. *Number of Repairs per Year.*—The number of repairs in a year for each of 10 such machines has been: 1, 6, 4, 4, 5, 5, 4, 2, 3, 6. The frequency distribution is:

Repairs	Probability	Repairs × Probability
6	.2	1.2
5	.2	1.0
4	.3	1.2
3	.1	.3
2	.1	.2
1	.1	.1
	Average	5.0

However, your engineer estimates the following probability for new machines based on his assessment of technical improvements:

Repairs	Probability	Repairs × Probability
5	.2	1.0
4	.4	1.6
3	.2	.6
2	.1	.2
1	.1	.1
	Expected number of repairs:	3.5

2. *MTTR (mean time to repair).*—The time consumed by the 40 repair jobs was recorded and a frequency distribution prepared:

Frequency	Hours Intervals	Mid-Point	Hours × Probability
2/40 = .05	0 – 2	1	.05
7/40 = .175	3 – 5	4	.70
18/40 = .45	6 – 8	7	3.15
10/40 = .25	9 – 11	10	2.50
3/40 = .075	12 – 14	13	.975
	Average time for repairs =		7.375 hours

Based on this information and his knowledge of product improvement your engineer estimates the following probability distribution:

Probability	Hours Mid-Point	Hours × Probability
4/40 = .10	1	.10
16/40 = .40	4	1.60
16/40 = .40	7	2.80
4/40 = .10	10	1.00
	Expected Average Hours:	5.5

Model: Expected annual labor cost = 3.5 × 5.5 × $10.00 =
Answer $192.50

Extrapolation Method

The extrapolation method is used for projecting costs on the basis of historical costs by projecting the trend. It is appropriate only where a trend exists. In Figure 4-1 two sets of costs are plotted against the time factor. In (a) a trend exists and extrapolation may be used; in (b) there is no trend discernible.

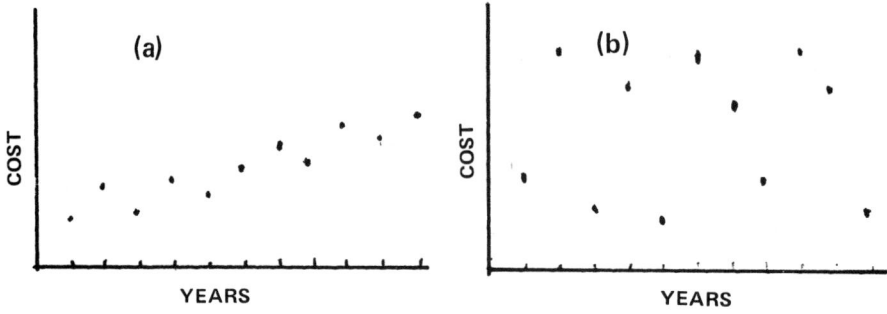

Figure 4-1.

For the benefit of readers with little mathematical background, this problem will first be solved in a nonmathematical fashion. Those familiar with statistics will understand the second solution and should find it to be a bit more accurate.

Figure 4-2 presents the Building Cost Index published by *Engineering News-Record,* covering the years 1967 to 1978. For the first method, plot the points on a graph with the cost index on the vertical axis and the years on the horizontal axis. Then draw a straight line through the points, endeavoring to obtain the best possible approximation of the trend. Extend this line to 1979 and 1980 and read the index figures for those years.

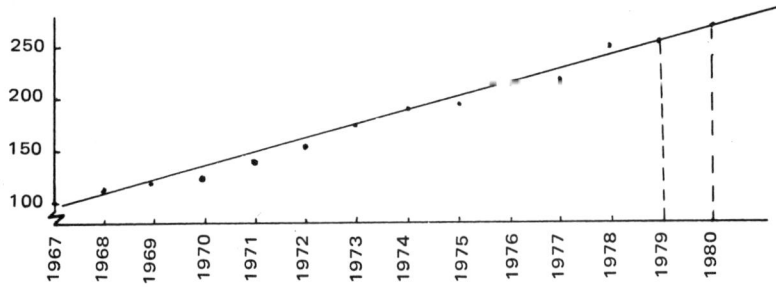

Figure 4-2. Building Cost Index

The values for 1979 and 1980, as determined by this extrapolation technique, are 1960 and 1975 respectively.

The second method to be used is linear regression. This is a technique for determining the equation of the straight line which is the best possible fit for the trend line. The linear regression can be solved using the formulas available in all basic statistics' texts or by use of a computer program. The latter method, using the Statistical Analysis System (SAS), is employed here. The program is as follows:

```
JOB CARD (280k needed)           For data cards:
/ / EXEC SAS                     Year in Cols. 1–4
/ / SYSIN DD *                   Index in Cols. 6–10
DATA BROWN;                      with decimal point
INPUT YEAR 1-4 PRICE 6-10;       in Col. 8
TREND=YEAR-1967;
CARDS;
data cards
PROC PRINT; VARIABLES YEAR PRICE;
PROC SYSREG; MODEL PRICE=YEAR;
MODEL PRICE=TREND;
PROC PLOT; PLOT PRICE*YEAR;
/ *
```

The printout reveals that the line crosses the Y axis at 90.45 and that the trend value is 13.432. That is,

$$Y = 90.45 + 13.432X$$

where Y is the vertical axis (cost index) and X is the horizontal axis (years). For the first year in the series, 1967, X is given the value zero; for 1968, X = 1, and so on. This equation represents the "best fit" line and it is used to calculate future values of the cost index thus:

Year	Equation		Index
1979	Y = 90.45 + 13.432(13)	=	265
1980	Y = 90.45 + 13.432(14)	=	278

Curve Fitting Method

Certain types of annual costs behave in a fashion which may be described mathematically. When plotted on a graph and a line drawn through them it may be that the line will be a straight line or a curve. If the line fits closely, the equation of the line may be used to describe the cost behavior and to estimate costs. An example of this type of analysis follows.

EXAMPLE 4-1

A firm is considering the purchase of a stamping machine similar to four others presently in use. In order to use life cycle costing the firm needs to forecast maintenance costs. Given the maintenance costs of the machines on hand, estimate the costs for a new machine.

Maintenance Costs			Years of Operation					
Machines	1	2	3	4	5	6	7	8
1	310	412	500	580	652	710	762	800
2	300	440	440	530	620	650	740	810
3	330	480	570	590	650	690	720	760
4	320	430	530	540	630	650	710	750

SOLUTION

Method 1: Obtain the yearly averages.

	1	2	3	4	5	6	7	8
answers:	315	438	510	560	638	675	733	780

Method 2: Obtain the equation of the curve and insert the years. The equation for the curve may be obtained by using the SAS program for the solution based on the model:

$$Y = A(1-e^{-Bx})$$

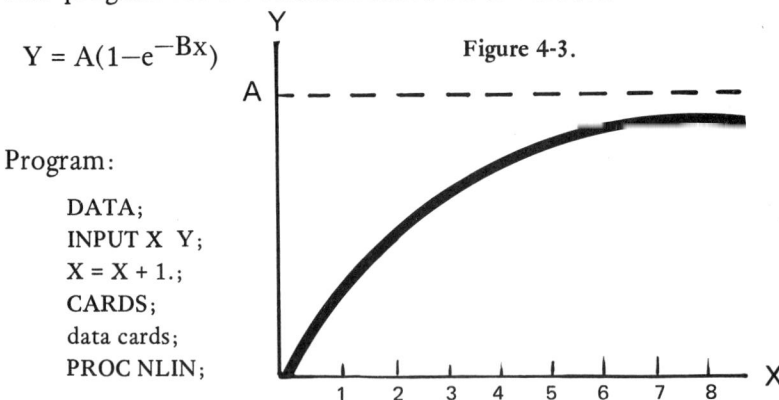

Figure 4-3.

Program:

```
DATA;
INPUT X Y;
X = X + 1.;
CARDS;
data cards;
PROC NLIN;
```

```
PARMS B0 = 760 TO 770 BY 1  B1 = .1 TO .9 BY .1;
MODEL Y = B0* (1-EXP(-B1*X));
DER.B0 = 1 - EXP(-B1*X);
DER.B1 = B0*X*EXP(-B1*X);
```

Equation: $Y = 898(1 - c^{-.209X})$

Now substitute the values of X to calculate Y:

X	1	2	3	4	5	6	7	8
Y	307	418	509	582	642	690	729	761

Note: Since the curve rises sharply from the 0 intercept, a better fit has been obtained by letting X = X + 1, that is by adding 1 to the year.

SENSITIVITY ANALYSIS

There is almost always some uncertainty in projecting costs regardless of whether they are variables, parameters, or constants. A true concern is what effect small changes in the specific costs will have on the total costs. The process of determining how much the LCC will change as a result of a change in one of the input factors, other things being held constant, is known as a "sensitivity analysis."

The first step in sensitivity analysis is to calculate the LCC for the *base case,* that is, for the most likely estimates. Then, for the various input factors of interest compute the LCC with small changes above and below the base case.

Illustration

$$LCC = wx + my + ez$$

where
  w = wage rate
  x = hours worked
  m = materials unit cost
  y = units of materials used
  e = energy cost per unit
  z = units of energy used

Let's assume for the base case LCC = $10,000. For 1.05m, LCC = $10,100 and for .95m, LCC = $9,900. For 1.05w, LCC = $10,500 and for .95w, LCC = $9,500. A 5% change in the materials cost per unit will produce a 1% change in LCC while a 5%

change in wage rate will cause a 5% change in LCC. (See Figure 4-4.).

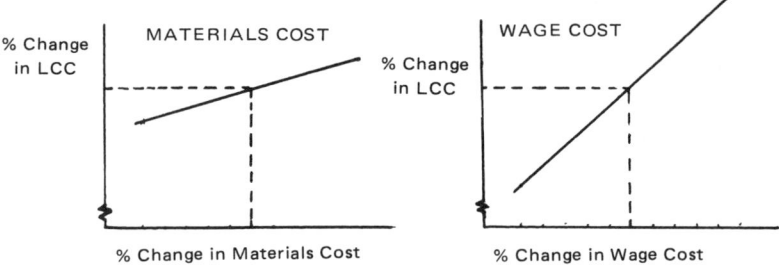

Figure 4-4.

Conclusion

The LCC is much more sensitive to a 5% change in wage rate than to a 5% change in materials cost.

Sensitivity analysis involving changes of several parameters can require tedious computation and use of a programmable calculator or a computer is recommended. Development of a computer program for a life cycle analysis is presented in a later chapter of this book.

EXPECTED VALUE ANALYSIS

It may be useful, in connection with sensitivity analysis, to go a step further by using "expected value analysis." Sensitivity analysis, in itself, may not be too helpful at times in pinpointing a "best" choice since its only purpose is to assay the effects of certain changes. Expected value analysis, however, does provide guidance of a specific nature which may be used in the decision process.

First, a word about the institutional framework for the analysis: If life cycle costing is performed on the basis of best estimates with the understanding that the project will go to the lowest bidder, sensitivity analysis and expected value analysis is irrelevant. This is often the case where a governmental body invites bids. Only where the decision-maker has the option to choose a project without regard to lowest LCC do these analytical techniques become useful. Even in those cases, however, they may fail to indicate a clear-cut solution. It can happen that competing projects are approximately equal.

As an illustration of the application of expected value analysis, take the case where two HVAC configurations are under consideration. The net present worth of the savings of each over the present system is computed for energy escalation rates from 4% to 11% (the expected range). All other factors are held constant. Results are given in Figure 4-5.

Energy Escalation Rate	HVAC Project A	HVAC Project B
4%	$1,491	$2,410
5%	1,984	2,656
6%	2,510	2,963
7%	3,072	3,200
8%	3,673	3,500
9%	4,314	3,821
10%	5,000	4,164
11%	5,736	4,532

Figure 4-5. Net Present Worth of Savings

The break-even point for A and B is between 7% and 8%. If there is a very high probability that the escalation rate will be 7% or less, B should be chosen. A should be chosen if an escalation rate of 8% or more is highly probable. If there is any doubt about the decision, a probability distribution should be created and the expected net present value of the savings computed. (A probability distribution is merely a listing of the possibilities and the probability of each. A simple example: Rain 70%, No rain 30%.)

To illustrate the concept, let's examine the process under each of two conditions or sets of probabilities. Begin by obtaining from management their estimates of the probabilities of experiencing each of the energy escalation rates in the likely range of 4% to 11%. These probabilities must add up to 1.00. The probability distributions are depicted in Figure 4-6.

These probability distributions will enable us to compare the solutions of the problem under two differing sets of conditions.

64   Life Cycle Costing: A Practical Guide for Energy Managers

Energy Escalation Rate	X Estimates	Y Estimates
4%	.02	.07
5%	.08	.11
6%	.15	.18
7%	.25	.29
8%	.25	.29
9%	.15	.06
10%	.08	
11%	.02	

Figure 4-6. Probability Distributions for Conditions X and Y

Next, multiply the probabilities by the savings at each level of escalation rate and total. The total is the expected net present worth of the savings for that project. The calculations for both X and Y estimates are given in Figure 4-7.

Energy Escalation Rate	A		Probability			B		Probability		
CONDITION X										
4%	1,491	X	.02	=	29.82	2,410	X	.02	=	48.20
5%	1,984	X	.08	=	158.72	2,656	X	.08	=	212.48
6%	2,510	X	.15	=	376.50	2,963	X	.15	=	444.45
7%	3,072	X	.25	=	768.00	3,200	X	.25	=	800.00
8%	3,673	X	.25	=	918.25	3,500	X	.25	=	875.00
9%	4,314	X	.15	=	647.10	3,821	X	.15	=	573.15
10%	5,000	X	.08	=	400.00	4,164	X	.08	=	333.12
11%	5,736	X	.02	=	114.72	4,532	X	.02	=	90.64
					3,413.11					3,377.04

Difference: $36.07 in favor of A.

CONDITION Y

4%	1,491	X	.07	=	104.37	2,410	X	.07	=	168.70
5%	1,984	X	.11	=	218.24	2,656	X	.11	=	292.16
6%	2,510	X	.18	=	451.80	2,963	X	.18	=	533.34
7%	3,072	X	.29	=	890.88	3,200	X	.29	=	928.00
8%	3,673	X	.29	=	1,065.17	3,500	X	.29	=	1,105.00
9%	4,314	X	.06	=	258.84	3,821	X	.06	=	229.26
					2,989.30					3,166.46

Difference: $177.16 in favor of B.

Figure 4-7. Expected Net Present Worth of Savings

It will be observed from Figure 4-7 that given the X estimates the expected net present worth of savings from Project A exceed those of B. For the Y estimates, the expected net present worth of B's savings exceed those of A. In both X and Y conditions the 7% and 8% energy escalation rates were the most probable and they were equiprobable. Only when the full probability distributions were applied did the decisions become apparent.

Finally, it should be clear that even though the probabilities associated with the various levels of energy escalation rates are only "guesstimates," the figures do represent an approximation of latent information which is relevant and which should not be overlooked. Do not throw away information.

SELF-STUDY PROBLEMS

Problem 4-1.—Below is a probability distribution of the annual escalation rate for electricity for the next 6 years. Also given is the breakeven period for a lighting installation figured on the basis of the indicated escalation rate. Assuming all other cost factors are constant, should the company adopt the project if its payback criterion is 2.2 years at the 90% probability level (that is, the company wants 2.2 years or less)?

Escalation Rate	Probability	Breakeven (Years)
6%	.10	2.4
7%	.20	2.3
8%	.40	2.2
9%	.20	2.1
10%	.10	2.0

Problem 4-2.—A school district is planning to construct a school. The following factors are some of the components of the LCC model. Classify them as constants, variables or parameters.

Shape of building
HVAC system
Size of classrooms
Number of classrooms
Energy inflation rate

Materials used
Area of plot
Costs of materials
Bathroom areas
Discount rate

Problem 4-3.—In the equipment repair problem above (Expert Judgment Method), we used a wage rate of $10 an hour and treated the rate as a constant. However, management is free to raise the rate. They are considering an increase in the rate in the expectation that the employees will work harder and thus reduce the MTTR. The MTTR probability distributions associated with each level of wage rate is given below. The expected number of repairs per year is 3.5. Perform the analysis to determine the optimal wage rate.

Rate $11 an hour	Probability	Hours Mid-Point
	.15	1
	.50	4
	.25	7
	.10	10
Rate $12 an hour	.20	1
	.55	4
	.15	7
	.10	10

5
Energy Cost Estimation

The escalation of energy costs, more than any other single factor, has advanced the use of life cycle costing and, therefore, it is understandable that a great many of the applications of life cycle costing are energy-related. Life cycle costing is an obvious technique for evaluating projects having continuing energy costs.

The purpose of this chapter is to familiarize the reader with some of the abbreviations, technical terminology, and formulas encountered in problems relating to energy consumption. The chapter may be passed over without interruption of thought flow but practitioners with particular interest in energy-saving applications should find it helpful.

Building-related projects receive most of the attention in this chapter since there are so many components of a building that have annual operating costs associated with energy. Buildings, especially those that are space conditioned year-round for temperature and humidity control, are quite energy intensive.

If a profile of the various costs of a typical $1,000,000 building were plotted over the life of that building, it might look something like the one shown in Figure 5-1. As is obvious from the figure, the operating costs are a major portion of the total costs of a building. Because the energy cost is such a significant amount of the total operating cost, it is important that the factors contributing to the energy cost be understood by the owner.

Familiarity with the characteristics of the energy-intensive components of a building is a must for building owners and

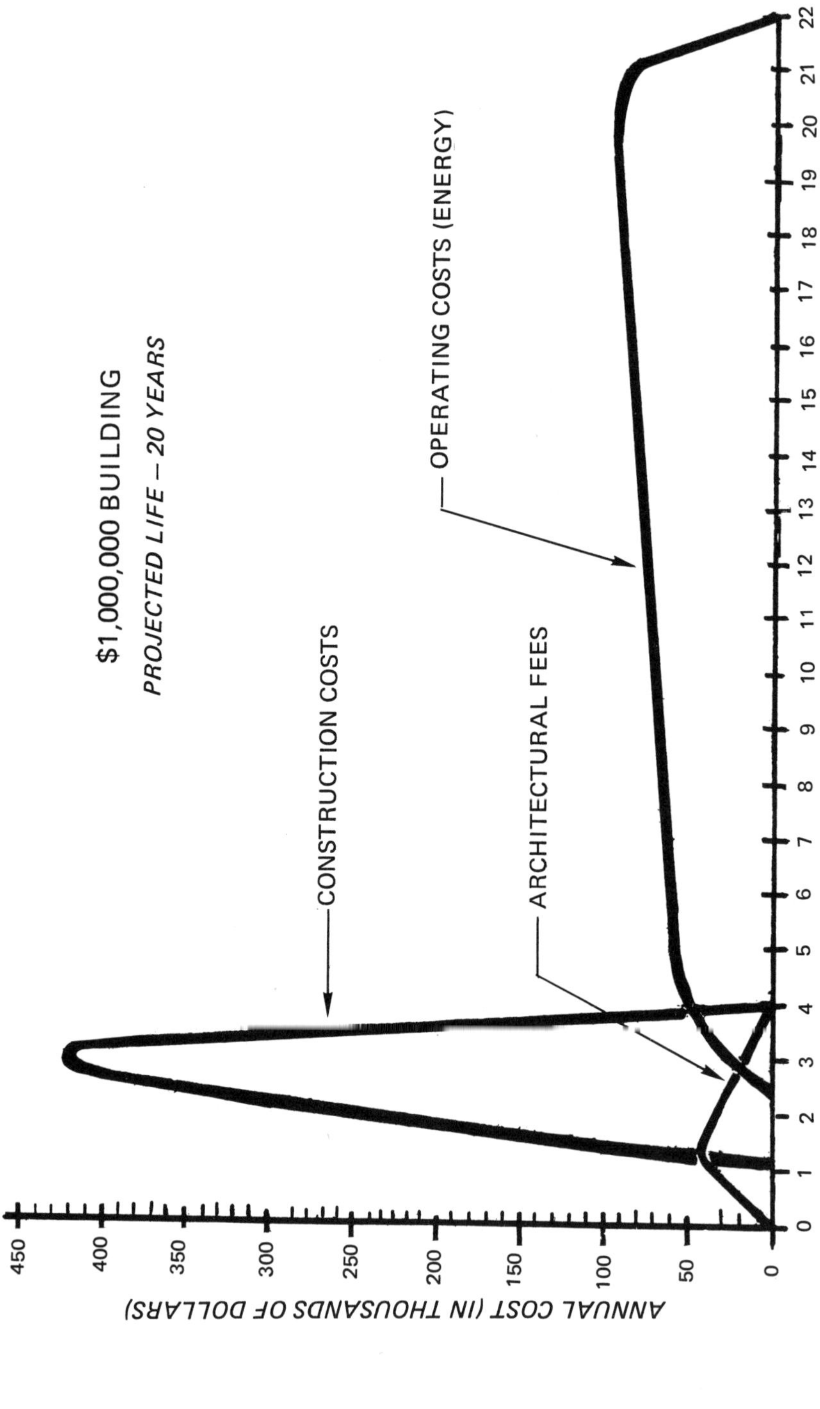

68 Life Cycle Costing: A Practical Guide for Energy Managers

Figure 5-1. Stages of Life Cycle Cost

managers. Some typical energy-intensive components that deserve attention are: walls and wall insulation, windows, pumps, motors, lighting, ventilating equipment, and doors, just to name a few. Energy cost calculations for these components and others are given in the following pages.

Table 5-1 provides a listing of abbreviations commonly used by professionals. Table 5-2 lists important energy values.

Table 5-1. Abbreviations

AMP	Amperage
AREA	Area under consideration; square feet
bbl	Barrel
Btu	British thermal unit
Bhp	Boiler horsepower
CFM	Cubic feet per minute
CF or cf	Cubic feet
d	Day
D.D.	Degree days
Eff.	Efficiency (ratio of energy or work output to the energy or work input)
FPM	Feet per minute
Fuel unit	Unit of purchasing energy; i.e., gal., KWH, CF, tons
Gal.	Gallon
Hp	Horsepower
hrs	Hours
I.A.	Inside air (refers to the air within a building)
L	Length in feet
KVA	Kilovolt amps; product of voltage and current. K = 1000
KW	Kilowatt
KWh	Kilowatt hours
MPH	Miles per hour
O.A.	Outside air (average outside air temperature may be obtained from the U.S. Weather Bureau)
PSI	Pound per square inch

Table 5-1. (continued)

R	Thermal resistance to heat flow; $\dfrac{\text{sq ft} \times \text{hr} \times {}^\circ\text{F}}{\text{Btu}}$
R_T	Total thermal resistance of composite structure
sq ft	Square foot
ΔT	Temperature difference (temperature difference between inside and outside air)
U	Thermal conductance of a composite structure; $\dfrac{\text{Btu}}{\text{sq ft} \times \text{hr} \times {}^\circ\text{F}} = \dfrac{1}{R_T} = U$
V	Voltage

Table 5-2. Important Energy Values

Btu — British thermal unit	= heat required to raise the temperature of 1 pound of water by 1 °F
Degree day (1)	= 65 °F minus mean temperature of the day.
Therm	= 100,000 Btu
1 U.S. gallon water	= 8.335 pounds at 60 °F
1 U.S. barrel (bbl)	= 42 U.S. gallons
1 Kilowatt-hour (KWH)	= 3,416 Btu (approx.) 1.341 horsepower-hour
1 horsepower-hour	= 2,545 Btu 0.746 KWH
1 gallon oil No. 2	= 138,000 Btu (approx.)
1 gallon oil No. 6	= 144,000 Btu (approx.)
1 cubic foot natural gas	= 1,000 Btu
1 ton coal (anthracite)	= 25 to 27 $\times$ 10^6 Btu
1 ton coal (bituminous)	= 20 to 27 $\times$ 10^6 Btu
1 pound propane or LP gas	= 21,500 Btu
1 pound steam @ 212 °F	= 970 Btu/lb
1 boiler horsepower (Bhp)	= 34.5 lbs steam/hr @ 212 °F = 33,472 Btu/hr = 9.810 KW
1 ton refrigeration	= 12,000 Btu/hr

Energy Cost Estimation

The reader should be aware that the Btu values given in Table 5-2 are averages. Btu content can vary, depending upon the source of supply. Bituminous coal from Western Pennsylvania, for example, may have a Btu content of 24×10^6 Btus per ton while bituminous coal from Wyoming might have a value of 18×10^6 Btus per ton. The analyst should know what kind of energy is being used and the Btu content of that energy source.

In this chapter basic equations are presented which can be used to determine energy costs of common building components. The equations given are not always meant to be used to obtain precise engineering results, but, rather, to approximate annual energy costs. For example, degree days are used in determining the heat loss of a building. A more exact measurement may be obtained by using hourly temperature differences for the year. However, the use of degree days will provide close approximation of the annual energy costs. In life cycle costing, where differences in alternative costs are of interest, approximations of this nature will generally provide fairly accurate results.

FORMULAS FOR COMPUTING BUILDING ENERGY COSTS

Annual Lighting Cost — ALC

$$\boxed{ALC = \frac{Watt \times hrs/yr}{1000} \times \$/KWH} \qquad \textit{Formula 5-1}$$

EXAMPLE

What does it cost to operate a 200-watt incandescent light bulb for 8 hours a day, 365 days a year if electricity cost is $.06/KWH?

SOLUTION

$$ALC = \frac{200 \times 8 \times 365}{1000} \times .06$$

$$ALC = \$35.04$$

COST REDUCTION IDEAS — LIGHTING

- Reduce excessive illumination
- Provide local switching
- Provide task lighting
- Replace inefficient lamps with high efficient lamps
- Utilize natural lighting when possible
- Clean fixtures and lamps
- Relamp on regular basis
- Use dimming equipment
- Install timers on light switches in little used areas
- Reduce exterior building and grounds illumination to minimum safe level
- Lower lighting fixtures in high ceiling areas
- Increase light reflectance of walls and ceilings
- Eliminate inefficient electric lamps from plant stocks and catalog

Transmission Heat Loss — THLC

$$THLC = \frac{U \times Area \times 24 \times DD}{Eff \text{ (of heating system)}} \times \frac{\$/Fuel\ Unit}{Btu/Fuel\ Unit}$$

Formula 5-2

EXAMPLE

Compute the wall heat loss given the following:

Energy Cost Estimation

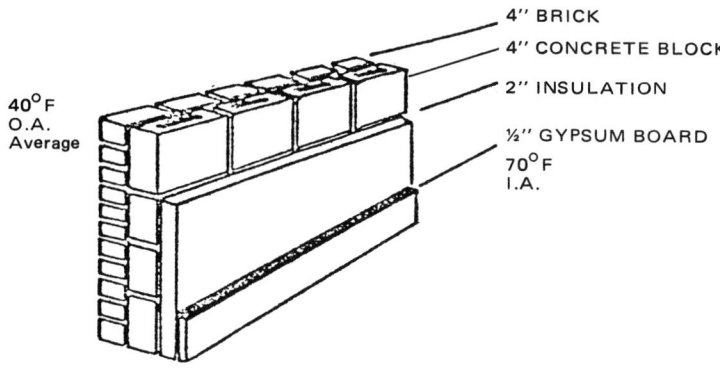

THERMAL RESISTANCES

Outside air film	0.17
4" brick	0.98
4" concrete block	0.63
2" insulation	8.70
½" gypsum board	0.45
Inside air film	0.68
R_T = Total Resistance	11.61

$$U = \frac{1}{R_T} = \frac{1}{11.61} = .09 \text{ Btu/hr, SF, }°F$$

(See ASHRAE Handbook for further details)

That is, 0.09 Btus are lost per hour per square foot of wall for each degree F between the temperature of the interior and exterior air.

A = 1000 sq ft
No. 2 fuel oil at 138,000 Btu gal.
Efficiency of heating system = 70%
$/Fuel Unit = $.65/gal.
DD = 5000

SOLUTION

$$\text{THLC} = \frac{.09 \times 1000 \times 24 \times 5000}{.70} \times \frac{.65}{138,000}$$

$$\text{THLC} = \$72.67$$

COST REDUCTION IDEAS — TRANSMISSION HEAT LOSS

- Add wall, ceiling and floor insulation where practical
- Set temperature back during unoccupied hours
- Install time clocks for night setback
- Reduce temperature in unused rooms
- Use radiant heater for spot heating rather than heating entire area
- Air condition only space in use
- Use heat pump for space conditioning instead of electric resistance heat

Transmission Heat Loss Costs (Windows) = THLC

$$\text{THLC} = \frac{U \times \text{Area} \times 24 \times DD}{\text{Eff (of heating system)}} \times \frac{\$/\text{Fuel Unit}}{\text{Btu}/\text{Fuel Unit}}$$

Formula 5-3

U factor is selected from a typical table such as the abbreviated one given below (reprinted with permission from the 1977 Fundamentals Volume, *ASHRAE Handbook and Product Directory*).

PART A—VERTICAL PANELS (EXTERIOR WINDOWS, SLIDING PATIO DOORS, AND PARTITIONS)— FLAT GLASS, GLASS BLOCK, AND PLASTIC SHEET

	Exterior		
Description	*Winter*	*Summer*	*Interior*
Flat Glass			
single glass	1.10	1.04	0.73
insulating glass—double			
0.1875-in. air space	0.62	0.65	0.51
0.25-in. air space	0.58	0.61	0.49
0.5-in. air space	0.49	0.56	0.46
0.5-in. air space, low emittance coating			
$e = 0.20$	0.32	0.38	0.32
$e = 0.40$	0.38	0.45	0.38
$e = 0.60$	0.43	0.51	0.42
insulating glass—triple			
0.25-in. air spaces	0.39	0.44	0.38
0.5-in. air spaces	0.31	0.39	0.30

(continued)

Energy Cost Estimation

Description	Exterior Winter	Exterior Summer	Interior
storm windows			
1-in. to 4-in. air space	0.50	0.50	0.44
Plastic Sheet			
single glazed			
0.125-in. thick	1.06	0.98	—
0.25-in. thick	0.96	0.89	—
0.5-in. thick	0.81	0.76	—
insulating unit—double			
0.25-in. air space	0.55	0.56	—
0.5-in. air space	0.43	0.45	—
Glass Block			
6 X 6 X 4 in. thick	0.60	0.57	0.46
8 X 8 X 4 in. thick	0.56	0.54	0.44
—with cavity divider	0.48	0.46	0.38
12 X 12 X 4 in. thick	0.52	0.50	0.41
—with cavity divider	0.44	0.42	0.36
12 X 12 X 2 in. thick	0.60	0.57	0.46

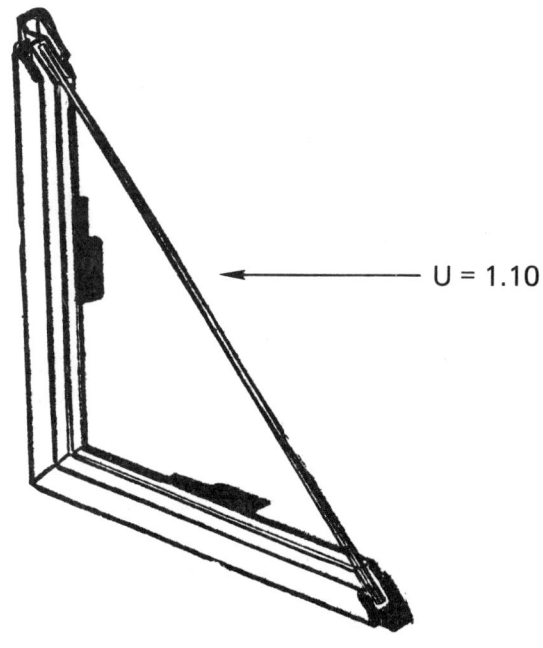

U = 1.10

EXAMPLE

What is the annual transmission heat loss cost for a 100-square-foot, single-pane window using the following data:

DD = 5000
Heating Source = No. 2 fuel oil at 138,000 Btu/gal. with a heating efficiency of 70%. Oil cost is $.65/gal.

SOLUTION

$$\text{THLC} = \frac{1.10 \times 100 \times 24 \times 5000}{.70} \times \frac{\$.65}{138,000}$$

$$\text{THLC} = \$88.82$$

**COST REDUCTION IDEAS —
WINDOW TRANSMISSION HEAT LOSS**

- Add storm windows
- Shades and drapes reduce heat loss at night
- Solar heat helps heat room
- Reduce window area on the north side of buildings
- Install insulating curtains
- Repair broken or cracked windows

Heating Infiltration Air Costs = HIAC

$$\boxed{\text{HIAC} = \frac{\text{CF} \times \text{L} \times 24 \times \text{DD}}{55.56 \times \text{Eff}} \times \frac{\$/\text{Fuel Unit}}{\text{Btu/Fuel Unit}}}$$

Formula 5-4

Cubic feet of infiltration air per foot of crack of window may be obtained from the following table. (Infiltration from door cracks may be obtained in a similar manner.)

INFILTRATION THROUGH WINDOWS
Cubic Feet of Air per Foot of Crack per Hour

	Wind Velocity, Miles per Hour					
Wood Windows	5	10	15	20	25	30
Around Frame						
Wood frame wall	2.2	6.2	10.8	16.6	23.0	30.2
Masonry wall, not calked	3.3	8.2	14.0	20.2	27.2	34.7
Masonry wall, calked	.5	1.5	2.6	3.8	4.8	5.8
Around Sash—Double Hung						
1/16″ Crack, plain	4.5	15.3	28.5	42.7	57.5	73.5
1/16″ Crack, weatherstripped	2.2	6.9	12.8	19.0	25.6	33.1
3/32″ Crack, plain	24.7	62.8	99.6	137.	176.	220.
3/32″ Crack, weatherstripped	3.8	12.7	23.3	34.9	47.5	61.2
Around Sash—Casement						
Plain double window	40.4	80.8	121.	162.	202.	242.
Weatherstripped double window	1.86	3.72	5.58	7.44	9.30	11.2
Metal Windows						
Around Frame						
Steel mullions, 3/64″ crack	8.	16.	28.	44.	63.	84.
Steel framing, 3/64″ crack	14.	30.	52.	76.	105.	133.
Around Sash—Double Hung						
Plain window	20.4	47.3	74.0	104.	137.	171.
Weatherstripped window	6.5	18.6	31.5	46.0	60.2	76.0
Around Ventilator—Casement						
Industrial pivoted, 1/16″ crack	52.	108.	180.	244.	310.	380.
Architectural projected, 3/64″ crack	24.	52.	88.	118.	155.	190.
Resident casement 1/32″ crack	12.	32.	54.	77.	102.	125.
Doors						
Plain door, 1/16″ crack, 1/16″ clearance	57.	114.	159.	227.	284.	341.
Plain door, 1/8″ crack, 1/16″ clearance	86.	171.	257.	342.	428.	514.
Weatherstripped door, all conditions	2.08	4.16	6.24	8.32	10.8	12.5

Source: *Fan Engineering*, Buffalo Forge Co., Fifth Edition.

EXAMPLE

A wooden 3 × 5 double hung window is installed in a masonry wall and caulked. A 1/16″ crack is assumed around the sash with a 15 mph wind velocity. The degree days = 5000 and the heating source is electric resistance heat with a Btu content of 3416 Btu/KWH at an efficiency of 100%. Determine heating infiltration cost per year. Electricity cost is $.06/KWH.

SOLUTION

Sash Crack Infiltration = CF × L = 28.5 × [(2 × 5) + (3 × 3)] = 541.5

Frame Infiltration = CF × L = 2.6 × [(2 × 5) + (2 × 3)] = 41.6

583.1

$$\text{HIAC} = \frac{583.1 \text{ CFM} \times 24 \text{ hr} \times 5000 \text{ DD}}{55.56 \times 1.0} \times \frac{\$.06/\text{KWH}}{3416 \text{ Btu/KWH}}$$

HIAC = $22.12/yr

COST REDUCTION IDEAS – INFILTRATION

- Caulk and seal around windows and doors
- Repair or install new outdoor air dampers
- Seal around switches and receptacles
- Insulate thoroughly around lighting fixtures
- Seal around all exterior openings
- Don't keep doorways open excessively
- Install storm windows and doors

Water Heating Costs – WHC

Btus used per year = gals./yr × 8.34 lbs/gal. × ΔT

$$\boxed{\text{WHC} = \frac{\text{Btus per year}}{\text{Eff.}} \times \frac{\$/\text{Fuel Unit}}{\text{Btu/Fuel Unit}}} \qquad \textit{Formula 5-5}$$

HOT
H_2O

Energy Cost Estimation

EXAMPLE

Compute the water heating cost given the following: 1000 gal./hr of water is needed for a manufacturing process. Water temperature is 190 °F supplied at 55 °F, 6 hrs/shift, 250 days/yr. Natural gas burned at 65% efficiency costing $3.00/1000 cf is used to heat the water.

SOLUTION

$$WHC = \frac{1000 \text{ gal./hr} \times 8.34 \text{ lb/gal.} \times (190\text{-}55) \,°F \times 6 \text{ hr/d} \times 250 \text{ d/yr}}{.65}$$

$$\times \frac{\$3.00/1000 \text{ cf}}{1000 \text{ Btu/cf}}$$

$$WHC = \$7795/\text{yr}$$

COST REDUCTION IDEAS – HOT WATER

- Reduce flow of water from shower heads
- Reduce hot water temperature
- Install point of use hot water heaters where higher temperature water is needed
- Restrict hot water at faucets to ¼ gpm
- Use cold water detergents in laundry
- Install self-closing faucets on hot water taps
- Install time clock to reduce temperature during unoccupied periods
- Increase efficiency of hot water generator
- Use cold water for clean up whenever possible
- Install separate heater for summer generation of domestic hot water

Heating Make-Up Air Cost – HMAC

$$\boxed{HMAC = \frac{1.08 \times CFM \times hr/d \times DD}{\text{Eff.}} \times \frac{\$/\text{Fuel Unit}}{\text{Btu/Fuel Unit}}}$$

Formula 5-6

This formula assumes no temperature setback.

80 Life Cycle Costing: A Practical Guide for Energy Managers

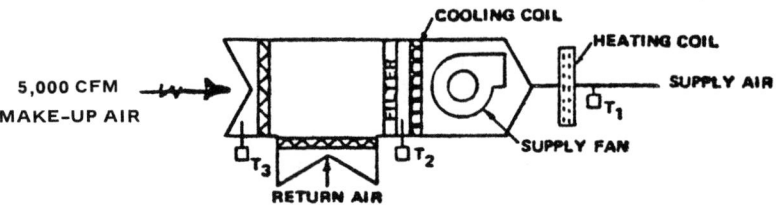

EXAMPLE

Compute the heating make-up air cost given the following: 5,000 CFM of outside air is heated 24 hrs/day to 70 °F. Degree Day = 5,200; eff. = 65%. No. 2 Fuel Oil, costing $.60/gal. is used.

SOLUTION

$$\text{HMAC} = \frac{1.08 \times 5000\,\text{CFM} \times 24 \times 5200\,\text{DD}}{.65} \times \frac{\$.60/\text{gal.}}{138{,}000\,\text{Btu/gal.}}$$

$$\text{HMAC} = \$4{,}508/\text{yr}$$

COST REDUCTION IDEAS — HEATING MAKE-UP

- Clean and replace filters regularly
- Reduce ventilation air to the required minimum
- Reduce room temperature to minimum standard
- Shut off air conditioner during the winter heating season
- Use direct air supply to exhaust hoods
- Close outdoor air dampers during warm-up or cool-down periods each day
- Adjust outside air damper to reduce outside air to the required minimum

Air Leakage of Dampers Cost — ALC
(during heating season)

If ventilation is required 24 hours a day, then:

$$\boxed{\text{ALC} = \frac{1.08 \times \text{CFM} \times \text{Leakage} \times 24 \times \text{DD}}{\text{Eff.}} \times \frac{\$/\text{Fuel Unit}}{\text{Btu/Fuel Unit}}}$$

Formula 5-7

Where: Leakage = $\dfrac{T_{RA} - T_{MA}}{T_{RA} - T_{CA}}$

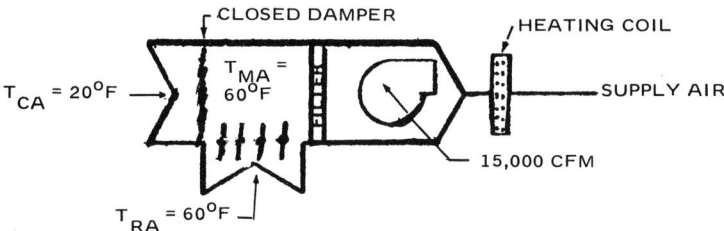

EXAMPLE

An outside air damper is in the closed position. The outside temperature, T_{CA}, is 20 °F, the return air temperature, T_{RA}, is 70 °F, the mixed air temperature, T_{MA}, is 60 °F. The CFM of the air handler is 15,000 and the degree days for the area are 5,000. Oil costs $.65/gal. and heating efficiency is 65%. What is the cost to heat the air due to damper leakage?

SOLUTION

$$\text{Leakage} = \frac{T_{RA} - T_{MA}}{T_{RA} - T_{CA}} = \frac{70 - 60}{70 - 20} = .20 \text{ or } 20\%$$

$$ALC = \frac{1.08 \times 15{,}000 \text{ CFM} \times .20 \times 24 \times 5000 \text{ DD}}{.65}$$

$$\times \frac{\$.65/\text{gal.}}{138{,}000 \text{ Btu/gal.}}$$

$$ALC = \$2817/\text{yr}$$

COST REDUCTION IDEAS – AIR LEAKAGE

- Install air seals at damper blade edges and ends
- Replace old damper with low leakage dampers

A.C. Motor Costs — ACMC

$$\text{ACMC} = \frac{.746 \times \text{HP} \times \text{hrs/yr} \times \$/\text{KWH}}{\text{Eff. Motor}}$$

Formula 5-8

EXAMPLE

A 25-horsepower A.C. motor is operated 2000 hours a year. Electrical energy costs $.06/KWH. Motor Efficiency is 85%. Compute the cost to operate the motor.

SOLUTION

$$\text{ACMC} = \frac{.746 \times 25 \times 2000 \times .06}{.85}$$

$$\text{ACMC} = \$2633$$

COST REDUCTION IDEAS — MOTORS
- Use most efficient type motor
- Use multiple speed motors or variable speed drives for variable pumps, blowers, and compressors
- Optimize motor size with load
- Shut off motors when not required

Energy Cost Estimation

Pump Cost – PC

$$PC = \frac{GPM \times Total\ Head \times hr/yr}{5{,}300 \times Pump\ Eff. \times Motor\ Eff.} \times \$/KWH$$ *Formula 5-9*

EXAMPLE

A 1000 gpm pump with a total head of 15 feet is operated 2000 hours a year. The pump efficiency is 65% and the motor efficiency is 85%. Electricity cost is $.06/KWH. What is the annual cost to operate the pump?

SOLUTION

$$PC = \frac{100 \times 15 \times 2000}{5300 \times .65 \times .85} \times .06$$

$$PC = \$61.47/yr$$

COST REDUCTION IDEAS – PUMP

- Choose pump capacity to match required demand
- Use most efficient pump type
- Operate pump only as required
- Replace over-size pumps with optimum size

ELECTRICAL POWER COST

There are several components of power costs that comprise an electric bill. The four basic components, excluding items such as local tax components, are as follows:

Energy Cost.—This component of the electrical bill is the cost for the amount of energy actually used. Electricity is measured in kilowatt hours, KWH, over a given period of time, normally one month.

Demand Cost.—Cost for the potential peak demand for power that the electric utility company must be prepared to deliver, but which may only be used by the customer for a short period of time during the billing period. Demand is usually determined by measuring the highest electric current requirement in any 15- to 30-minute period within each month. Demand is expressed in KW. (Some utilities express demand in KVA, Kilovolt Amperes).

Fuel Adjustment Cost.—This charge reflects increases in the electric utility's fuel costs. Fuel adjustment costs are expressed in $/KWH.

Power Factor.—Power factor of an electrical load is the ratio of the true power in watts (measured with a wattmeter) to the apparent power (obtained from the product of the voltage and current of the load). Often electrical utilities charge a penalty to a customer with a low power factor. With a low power factor, inductive resistive equipment draws a certain amount of unusable power over and above the total watt-hour meter power. Additional generating capacity must be supplied to meet this demand caused by low power factor.

The sum total of these components will amount to the total electrical bill.

Electrical Cost	=	Energy Cost	+	Demand Cost	+	Fuel Adj. Cost	+	Power Factor Penalty Cost

Formula 5-10

Energy Cost Estimation

$$\text{Electrical Cost} = \text{KWH}\left(\frac{\$}{\text{KWH}}\right) + \text{KW}\left(\frac{\$}{\text{KW}}\right) + \text{KWH}\left(\frac{\$}{\text{KWH}}\right) + \left(\% \times \text{demand} \times \frac{\$}{\text{KW}}\right)$$

EXAMPLE

An industrial user has received the monthly electrical bill which lists the electrical consumption as follows:

Energy Used	150,000 KWH
Maximum Demand	1,000 KW
Energy Cost	$.04/KWH
Demand Cost	$4/KW
Fuel Adjustment Cost	$.0095/KWH
*Power Factor Penalty = 10%	½% of demand for every 1% below Power Factor × cost/KW

What is the user's total electric bill?

SOLUTION

$$\text{Electrical Cost} = \left(150{,}000 \text{ KWH} \times \frac{\$.04}{\text{KWH}}\right) + \left(1000 \text{ KW} \times \frac{\$4.00}{\text{KW}}\right) + \left(150{,}000 \text{ KWH} \times \frac{\$.0095}{\text{KWH}}\right) + \left(.05 \times 1000 \text{ KW} \times \frac{\$4.00}{\text{KW}}\right)$$

$$= \$6000 + \$4000 + \$1425 + \$200$$

Electrical Cost = $11,625

From the above, it is readily seen that the demand cost is a substantial part of the total electrical bill. Therefore, it behooves the knowledgeable owner to try to reduce this component cost to its minimum value.

---

*This component value differs with different utilities.

Demand Cost — DC

$$DC = KW \times \$/KW$$

Formula 5-11

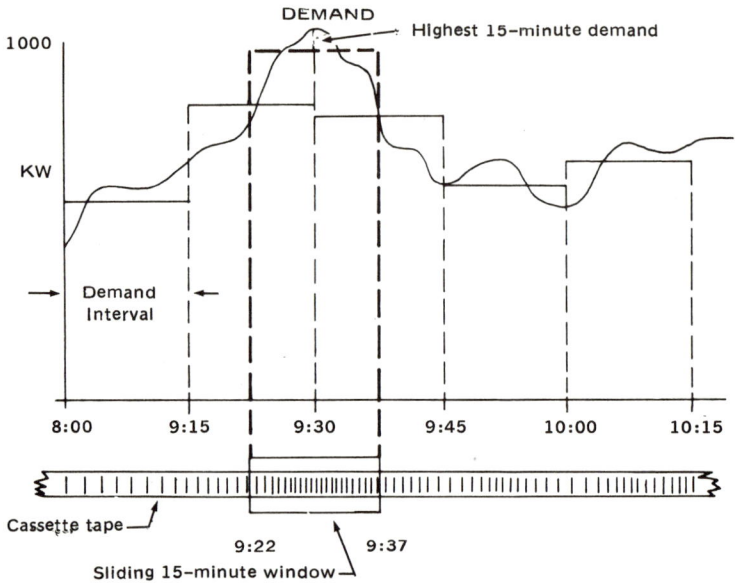

EXAMPLE

A peak demand of 1000 KW is recorded for one month's bill. What is the cost just for demand if the demand billing is as shown below?

Demand	Charge/KW
0 – 25 KW	No charge
26 – 100 KW	$4.75
over 100 KW	$4.50

SOLUTION

DC = (25 × 0) + (75 × $4.75) + (900 × $4.50)

DC = $4,406.25

COST REDUCTION IDEAS — DEMAND

- Heat domestic hot water during off-peak hours
- Schedule loads to minimize demand charge
- Use energy recovery systems to reduce load
- Schedule different start-up times for different operations

Uninsulated Steam Line Energy Cost — USLC

$$\text{USLC} = \frac{\text{Btu/L} \times \text{L} \times \text{hrs/yr}}{\text{Eff.}} \times \frac{\$/\text{Fuel Unit}}{\text{Btu/Fuel Unit}} \quad \textit{Formula 5-12}$$

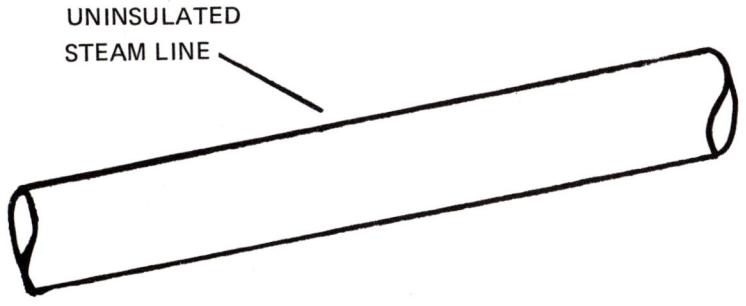

UNINSULATED STEAM LINE

The following abbreviated table lists the loss in Btu per hour from several sizes of uninsulated pipe with steam at the conditions listed.

Pipe Diameter	Steam Conditions		
	5 PSI 240°F	165 PSI 540°F	400 PSI 720°F
	Loss in Btu/Hr/Foot of Pipe		
½"	109	492	895
1"	164	747	1368
2"	282	1299	2397
4"	507	2370	4410

Source: *Energy Management Guide*, Pennsylvania Power & Light Co.

88   Life Cycle Costing: A Practical Guide for Energy Managers

EXAMPLE

A 4-inch uninsulated pipe carries 400-pound steam at 720°F 100 feet through the plant for a full 8760 hours per year. Coal (bituminous) at 26×10^6 Btu and a boiler efficiency of 70% are used. Coal cost is $30 per ton. Compute the annual loss due to the uninsulated steam line.

SOLUTION

Energy loss due to the uninsulated pipe would cost:

$$USLC = \frac{4410 \times 100 \times 8760}{.70} \times \frac{30}{26 \times 10^6}$$

$$USLC = \$6368$$

COST REDUCTION IDEAS — STEAM LINE

- Repair insulation on steam lines
- Repair insulation on condensate lines
- Insulate condensate storage tanks
- Turn off steam tracing during mild weather
- Clean steam coils in processing tanks
- Return 100% condensate to boiler
- Use minimum steam operating pressure
- Use optimum thickness insulation

Steam Leak Energy Cost — SLC

$$\boxed{SLC = \frac{\text{lbs/hr} \times \text{hrs/yr} \times 970 \text{ Btu/lb}}{\text{Eff.}} \times \frac{\$/\text{Fuel Unit}}{\text{Btu/Fuel Unit}}}$$

Formula 5-13

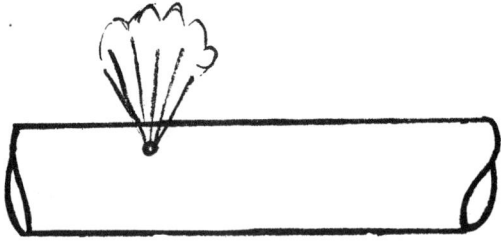

Energy Cost Estimation

The table below lists the approximate losses of steam in pounds per hour.

Hole Dia Inches	Flow Rate in Pounds Per Hour		
	Approximate Steam Pressure PSI		
	5	165	400
1/32	1	4	8
1/16	2	16	35
1/8	8	66	140
1/4	35	265	562
3/8	78	595	1266
1/2	139	1059	2250
5/8	217	1654	3515
3/4	313	2382	5062
7/8	426	3243	6890
1	556	4235	8999

Source: *Energy Management Guide*, Pennsylvania Power & Light Co.

EXAMPLE

A quarter-inch hole in a 400-pound-per-square-inch steam line will lose 562 pounds of steam per hour. Coal (bituminous) at 26×10^6 Btu/lb and a cost of $30/ton is used. The system efficiency is 60%. Compute the steam leak energy cost.

SOLUTION

$$\text{SLC} = \frac{562 \times 8760 \times 970}{.60} \times \frac{\$30}{26 \times 10^6}$$

$$\text{SLC} = \$9184^*$$

COST REDUCTION IDEAS – STEAM LINE

- Repair all steam line leaks
- Repair condensate line leaks
- Repair underground condensate line leaks
- Remove unneeded service lines to eliminate potential leaks

---

*Make-up water and sewage cost associated with leak should also be included for more accurate cost.

Leaky Steam Trap Costs — LSTC

$$\text{LSTC} = \frac{\text{Btu/1000} \times \text{hrs/yr}}{\text{Eff.}} \times \frac{\$/\text{Fuel Unit}}{\text{Btu/Fuel Unit}}$$

Formula 5-14

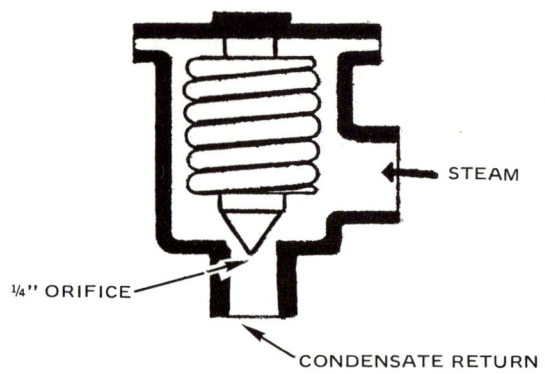

The table below illustrates the Btus lost per 1000 hours. (A steam trap fails in the "valve open" position.)

Losses Due to Steam Leaks Million Btu per Thousand Hours			
Steam Pressure	Orifice Size		
PSI	1/8"	1/4"	3/8"
20	12.3	47.9	79.9
50	30.8	120	200
100	61.6	240	400

Source: "Energy Management Guide for Light Industry and Commerce," NBS 1976

EXAMPLE

A steam trap with a ¼" orifice on a 20-psi steam is stuck open 24 hrs/day, 365 days/yr. No. 2 fuel oil is the source of energy to make the steam and costs $.80/gal. The efficiency of the system is 65%. Determine the annual cost due to the orifice leak.

SOLUTION

$$\text{LSTC} = \frac{47.9 \times 10^6/1000 \times 24 \text{ hrs} \times 365 \text{ d/yr}}{.65} \times \frac{\$.80/\text{gal.}}{138,000 \text{ Btu/gal.}}$$

$$\text{LSTC} = \$3742/\text{yr}$$

COST REDUCTION IDEAS – STEAM TRAPS

- Replace leaking traps with new or cartridge
- Conduct regular preventive maintenance

Open Door Cost (Loading Docks) – ODC

$$\text{ODC} = \frac{1.08 \times \text{FPM} \times \text{Area} \times \text{hrs} \times \Delta T}{\text{Eff.}} \times \frac{\$/\text{Fuel Unit}}{\text{Btu/Fuel Unit}}$$

Formula 5-15

OR

$$\text{ODC} = \frac{5,280 \times \text{mph} \times \text{Area} \times \text{hrs} \times \Delta T}{55.56 \times \text{Eff.}} \times \frac{\$/\text{Fuel Unit}}{\text{Btu/Fuel Unit}} \ *$$

Formula 5-16

*This formula is used when wind velocity is expressed in miles per hour.

EXAMPLE

A large manufacturering plant has an 8- × 10-foot loading door which opens to the outside. The door is kept open 2 hours per day during the heating season. The building temperature is heated to 60 °F from October to April for a total of 140 days per year. The Weather Bureau data indicates the average outside temperature for this area of the United States is 42 °F. The air flow velocity through the open door is 5 mph. How much does it cost to keep this door open if steam is used to heat the building? The appropriate efficiency for the heating system is 70%. Cost of the steam is $5.00/1,000 lbs.

SOLUTION

$$ODC = \frac{5{,}280 \times 5 \times (8 \times 10) \times 2 \times 140 \times (60 - 42)}{55.56 \times .70} \times \frac{\$5.00/1{,}000}{970}$$

$$ODC = \$1{,}410.79$$

COST REDUCTION IDEAS — LOADING DOCK

- Keep doors open a minimum amount of time
- Install air curtain
- Install plastic strip curtain
- Install vestibules
- Keep doors to minimum size required
- Install loading doors on leeward side of building

6
Service Life

Important to life cycle costing analysis is the decision of the service life, that is, the planning horizon to be used. Estimation is inevitable, but to the greatest extent possible, the estimation should be based on information. This chapter will deal with service life.

PLANNING PERIOD

The shorter the expected life of an asset, the more important is the choice of the planning period. This may be demonstrated by calculating the difference in average annual cost of two assets having initial costs of $1,050,000 and $1,000,000, using a 10% discount rate.

Initial Cost	Initial Cost Converted to Average Annual Cost			
	25 Years	30 Years	35 Years	40 Years
$1,050,000	$115,679	$111,384	$108,875	$107,373
1,000,000	110,170	106,080	104,690	102,260
	5,509	5,304	5,185	5,113
Difference	$205		$72	

If a life of 30 years were used in planning and the actual life, as foreseen in an infallible crystal ball, were 40 years, the error in average annual cost would be only $72. However, if a life of 25 years were used instead of 30, the error would be $205, or almost triple the 35-to-40 year difference.

The magnitude of error which may be experienced from choosing an incorrect service life on the short end of the planning spectrum may be vividly demonstrated by an illustration where choice is dictated by company policy on payback period.

EXAMPLE

An energy-conserving installation costing $12,000 is expected to save $3,000 per year for 10 years. The company requires a 3-year payback. Using a 10% discount rate find the life cycle cost for a 3-year life (to conform with the payback period) and a 10-year life.

SOLUTION

3-year life
Initial Cost	$12,000
Savings $3,000 (UPW, 3, 10%)	(7,640)
Life cycle cost =	$ 4,540

10-year life
Initial cost	$12,000
Savings $3,000 (UPW, 10, 10%)	(18,434)
Life cycle cost =	$ (6,434)

On the basis of a 3-year life, the savings fall short of repaying initial cost by $4,540. Over a 10-year life, the savings not only pay for initial cost, but also provide a net gain in present worth figures of $6,434. The use of payback period as a service life can lead to egregious errors.

Service life, or the planning horizon, is not necessarily identical with expected active life. It is, in the case of a product which is subject to sudden deterioration such as light bulb, but it may not be for products which deteriorate over a period of years. The difference between the two lives is that service life terminates at the point in time when the average annual costs of ownership and operation are at a minimum while active life ends when the product is no longer operable.

DETERMINING SERVICE LIFE

To determine service life, first draw up a trade-in schedule and a maintenance/replacement schedule. The trade-in schedule will list estimated resale values at the end of each year for the active life of the asset and the M/R schedule will list the estimated total annual maintenance and repair costs for the same period. These schedules are then used to calculate total average annual costs for each time period. (See Figure 6-1.) The service life is the year when the average annual costs are at a minimum and is the time to be used in planning purchases of equipment. In the calculations, costs which are constant throughout the active life of the asset may be excluded since they will have no bearing on the results.

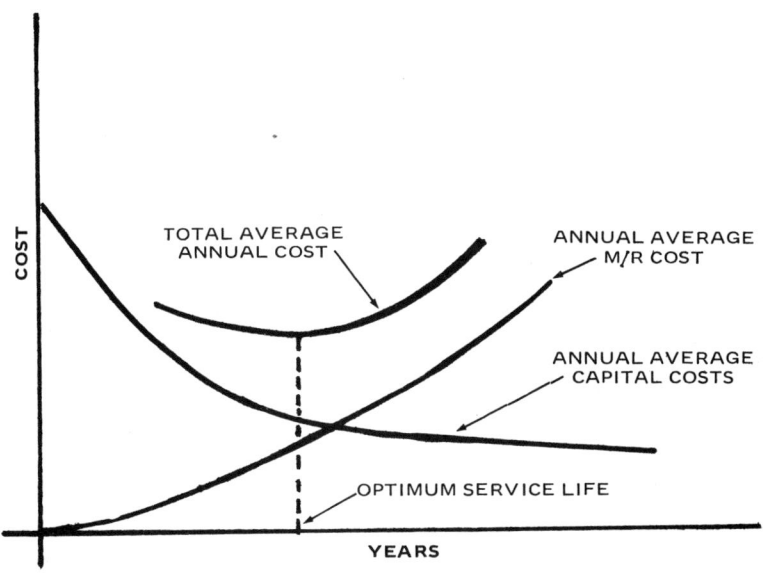

Figure 6-1

EXAMPLE

A machine costing $55,000 and expected to have an active life of 10 years has estimated trade-in and M/R schedules as set forth below. The discount rate will be 10%.

Year	Trade-In Schedule	M/R Schedule
1	$45,000	$ 1,000
2	36,000	1,000
3	28,000	2,500
4	21,000	3,000
5	15,000	3,500

(continued)

Year	Trade-In Schedule	M/R Schedule
6	$10,000	$ 4,500
7	6,000	9,500
8	3,000	10,000
9	1,000	11,800
10	0	12,500

(a) Determine the appropriate service life.
(b) Will the total life cycle cost be at a minimum in the service life year?
(c) How can a comparison be made between this machine and an alternative with a different service life?

SOLUTION

Year 1

Average annual capital cost =
$\bigl(55{,}000 - 45{,}000\,(\text{SPW};\,1,\,10\%)\bigr) \times (\text{UCR};\,1,\,10\%) =$ $15,500
Average Annual M/R cost = <u>1,000</u>
Total annual average cost = $16,500

Year 2

Average annual capital cost =
$\bigl(55{,}000 - 36{,}000\,(\text{SPW};\,2,\,10\%)\bigr) \times (\text{UCR},\,2,\,10\%) =$ $14,548
Average annual M/R cost =
Cumulative P. W. of prior M/R costs + P. W.
  of Year 2. M/R cost is $1,735.54
Therefore, $1,735.54 (UCR; 2, 10%) = <u>1,000</u>
Total annual average cost = $15,548

Year 3

Average annual capital cost =
$\bigl(55{,}000 - 28{,}000\,(\text{SPW};\,3,\,10\%)\bigr) \times (\text{UCR};\,3,\,10\%) =$ $13,657
Average annual M/R cost =
$\bigl(1{,}735.54 + 2{,}500 \times (\text{SPW},\,3;\,10\%)\bigr) \times$
  (UCR; 3, 10%) = <u>1,453</u>
Total annual average cost = $15,110

Summary

Year	Average Annual Capital Costs	Average Annual M/R Costs	Total Average Annual Costs
1	$15,500	$1,000	$16,500
2	14,548	1,000	15,548
3	13,657	1,453	15,110
4	12,826	1,786	14,612
5	12,052	2,067	14,119
6	11,332	2,382	13,714
7	10,665	3,133	13,798
8	10,047	3,733	13,780
9	9,477	4,327	13,804
10	8,951	4,840	13,791

ANSWER

(a) The minimum is at 6 years and therefore this is the service life. In other words, if the machine is replaced every 6 years, the average annual cost over a long period of time will be $13,714. If replaced at any other interval, the average annual cost will be higher.

(b) The total life cycle cost will be at a minimum in the first year and will increase each year since the costs are cumulative.

(c) Determine the average annual cost of each machine over its service life. The economic choice is the machine with the lower total average annual cost.

SELF-STUDY PROBLEM

Problem 6-1

A contractor finds that his loaders generally provide 8 years of service before breakdowns seriously interfere with operations. To find out what the service life should be, he has drawn up the following schedules. A loader costs $80,000 and his discount rate is 15%.

Year	Trade-In Schedule	M/R Schedule
1	$60,000	Warranty
2	50,000	$ 3,000
3	40,000	4,000
4	35,000	6,000
5	30,000	9,000
6	25,000	14,000
7	20,000	15,000
8	15,000	18,000

(To save time: The total average annual costs decline over the first four years, at which time they total $23,929, and the present worth of the total M/R costs at the end of year 4 is $8,329.01.)

7
Taxes and Depreciation

TAX CONSIDERATIONS

Organizations subject to federal and state income taxes must make allowances for such taxes in LCC analysis. As will be seen, taxes may be a critical factor in the decision among several alternatives.

Since tax-deductible expenses reduce the income subject to taxes, they produce a savings. Therefore, the expenses to be incurred from a project should be stated in the LCC analysis as their after-tax equivalents. In other words, a company paying 46% of its income in income taxes would incur expenses at only 54% of their value. The formula is: (100%−Firm's tax rate) X Expenses. The tax rates prescribed by the Revenue Act of 1978 are as follows:

Taxable Income	Tax Rate
$0 to $25,000	17%
$25,000 to $50,000	20%
$50,000 to $75,000	30%
$75,000 to $100,000	40%
Over $100,000	46%

Expenses relevant to LCC which are tax-deductible for federal income tax purposes include maintenance and operating costs, insurance, property taxes, state taxes. Interest expense is also deductible, but it should not be included since it is taken into consideration when the discount rate is determined.

A tax credit is a direct credit against regular tax liability. Details relevant to the 10% investment tax credit may be found in Appendix D. Also in Appendix D may be found information pertaining to the additional 10% tax credit which may be obtained for certain types of energy properties and the tax credits for home energy conservation. For full treatment of tax credits the reader is referred to the Internal Revenue Code as amended.

DEPRECIATION

Organizations which must pay federal income tax on earnings are permitted to amortize the cost of an asset over its life span as a charge against earnings. The original outlay is an expense, but the expense can only be charged off for tax purposes over the useful life of the asset. The term "depreciation," as used in this text, refers to the expensing of cost over the asset life and is not to be construed as synonymous with physical deterioration.

The most common method of accounting for depreciation is the straight-line method. The total amount which may be expensed by depreciation is prorated evenly over the useful life of the asset. For example, a warehouse costing $600,000 with useful life of 60 years would charge off $10,000 a year as depreciation expense. If a salvage value of $60,000 were expected, it would be deducted from cost prior to the division, i.e. $540,000 ÷ 60 = $9,000.

In life cycle analysis, the depreciation is not included as an expense since the original outlay is included as initial cost and depreciation would be duplicating it. However, there is a tax benefit to be derived from the depreciation expense and this benefit should be deducted from the annual costs. The formula used to calculate the tax benefit is:

$$\text{Firm's Tax Rate} \times \text{Depreciation Expense}$$

EXAMPLE

A firm pays 45% of its income in federal income taxes. An air-conditioning unit costing $30,000 will be depreciated by the straight-line method over 15 years. What is the annual tax benefit?

SOLUTION

$$45\% \times \$2{,}000 = \$900$$

ACCELERATED DEPRECIATION

The Internal Revenue Code allows accelerated depreciation for qualified assets having a useful life of at least 3 years. Accelerated depreciation permits earlier recovery of the depreciation tax benefits and is advantageous to firms seeking faster payback of outlays. Two common methods of accelerating depreciation are Sum-of-the-Years Digits (SYD) and Declining Balance (DB).

Under the Sum-of-the-Years Digits method, the amount to be depreciated (cost less expected salvage) is multiplied by a smaller fraction in each succeeding year. The numerator of the fraction is the number of remaining years of useful life and the denominator is the sum of the years digits.

That is,

Year 1 Depreciation: $\dfrac{n}{1 + 2 + \ldots + n}$

Year 2 Depreciation: $\dfrac{n - 1}{1 + 2 + \ldots + n}$

etc.

EXAMPLE

(a) SYD on project costing $15,000, life 5 years.

Year 1	5/15 × 15,000	=	5,000
Year 2	4/15 × 15,000	=	4,000
Year 3	3/15 × 15,000	=	3,000
Year 4	2/15 × 15,000	=	2,000
Year 5	1/15 × 15,000	=	1,000
			15,000 Total

(b) SYD on project costing $17,000, salvage value of $2,000, and life of 5 years.

Calculations are the same as above since the amount to be depreciated is cost less salvage.

The Declining-Balance method provides the greatest depreciation in the first year and declining amounts in the succeeding years. Unlike SYD, DB applies the same rate each year, but the amount to which the rate is applied is the cost minus previously deducted depreciation. Also, unlike SYD, salvage value is not deducted from the cost basis. The maximum rate which may be applied under the Declining-Balance method is twice the straight-line rate.

EXAMPLE

Calculate the annual depreciation for the first three years on an asset which cost $15,000 and is expected to have a life of 5 years. Use the maximum rate.

			Depreciation	Balance
Year 1	2/5 × 15,000	=	6,000	9,000
Year 2	2/5 × 9,000	=	3,600	5,400
Year 3	2/5 × 5,400	=	2,160	3,240

As the asset nears the end of its useful life, the taxpayer faces a decision. The taxpayer has the option of switching to straight-line and may wish to do so in order to avoid an increase in depreciation in the last year of useful life. This situation exists in Year 4 for the above asset. If the switch is not made, the depreciation is as follows:

			Depreciation	Balance
Year 4	2/5 × 3,240	=	1,296	1,944
Year 5	Remainder	=	1,944	0

If the switch is made, the depreciation is as follows:

			Depreciation	Balance
Year 4	1/2 × 3,240	=	1,620	1,620
Year 5	1/2 × 3,240	=	1,620	0

Although salvage is not deducted from cost for the initial basis under the DB method, it does play a role in the process in that the asset may not be depreciated below salvage value. In the above problem, for example, had the salvage been $1,000 the depreciation could have been:

	Depreciation	Balance
Year 4 2 × 3,240 =	1,296	1,944
Year 5 Remainder =	944	1,000

Double-Declining Balance, that is, the use of 200% of the straight-line rate, is permitted for tangible real property used for manufacturing, extraction, production, the furnishing of transportation, communication, electrical energy, gas, water and sewage disposal, for elevators and escalators, and for new residential rental property meeting requirements. Lower percentages are available for other purposes. The Internal Revenue Code should be consulted for further details.

Table 7-1 demonstrates the application of the three depreciation methods to the purchase of a machine costing $8,000, having an expected life of 8 years, and no salvage value. The depreciation recapture schedule is depicted in Figure 7-1.

Table 7-1

Year	Straight-Line		Double-Declining Balance		Sum-of-Years Digits	
	Depreciation	Cumulative Depreciation	Depreciation	Cumulative Depreciation	Depreciation	Cumulative Depreciation
1	1000	1000	2000	2000	1778	1778
2	1000	2000	1500	3500	1556	3334
3	1000	3000	1125	4625	1333	4667
4	1000	4000	844	5469	1111	5778
5	1000	5000	633	6102	889	6667
6	1000	6000	633*	6735	667	7334
7	1000	7000	633	7368	444	7778
8	1000	8000	632	8000	222	8000

*Switch to straight-line method.

ADR SYSTEM

The Asset Depreciation Range (ADR) system may be used for electing useful lives for eligible property placed in service after 1970. Eligible property is tangible personal and real property for which there is in effect an asset guideline class and an asset guideline period. For example, lightweight general purpose

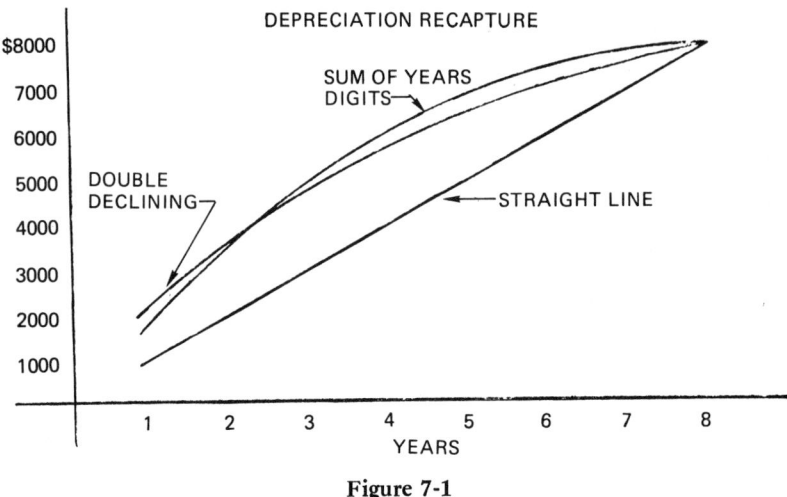

Figure 7-1

trucks have an ADR of 3 to 5 years. The taxpayer has the option of depreciating over a period of 3, 4, or 5 years.

Advantages of using the ADR system are:

(1) The useful lives are generally shorter than otherwise allowed.

(2) The salvage value will not be increased by IRS unless the difference between your estimate and theirs is 10% of cost.

(3) Salvage value is not considered in computing annual depreciation, but must be estimated because depreciation below salvage value is not permitted.

The ADR system may not be used where a depreciation method other than straight-line, SYD, or DB is used. Nor may it be used where special amortization is claimed, as for a certified pollution control facility.

AFTER-TAX COSTS

The reader is reminded that life cycle costing is applied principally to nonrevenue producing projects. Projects which produce revenue should be analyzed by the traditional capital budgeting methods since revenue cannot be ignored in the comparison of projects.

Also it should be clear that life cycle costs and average annual costs cannot be used by taxpayers for budgeting purposes since the costs are expressed on an after-tax basis. For planning, these costs should be returned to a pre-tax basis by dividing by (1 − tax rate).

SELF-STUDY PROBLEMS

Problem 7-1.—Determine the annual depreciation on an asset costing $45,000, with a life of 6 years and expected salvage value of $3,000. Use the straight-line, sum-of-years digits and double-declining-balance methods. Do not use ADR.

Problem 7-2.—Compare two HVAC systems given the following information. The planning horizon for both is 30 years and the firm's discount rate is 10%.

	System A	System B
Initial cost	$120,000	$90,000
Annual operating and maintenance costs	8,000	10,000
Salvage	0	0

Property tax rate is $80 per $1,000 of assessed valuation; assessed valuation is 40% of original cost. The firm's income tax rate is 40%. Straight-line depreciation is used.

SOLUTION

Present Worth	System A	System B
Initial cost	$120,000	
Operating and maintenance costs	45,249	
Property tax	21,720	
Depreciation tax benefit	(15,083)	
P. W. Totals	$171,886	

System A

Operating and
  maintenance costs: $ 8,000 × .60 × 9.42691 = $45,249

Property tax: $\dfrac{\$120,000}{1,000} \times \$80 \times .40 = \$3,840/\text{yr}$

$\$3,840 \times .60 \times 9.42691 = \$21,720$

Depreciation tax benefit: $\dfrac{120,000}{30} = \$4,000$

$\$4,000 \times .40 \times 9.42691 = \$15,083$

8
Lease or Buy

Ownership of an asset is not always the most economical way of obtaining use of the asset. Indeed, many of the assets used in our daily lives are not owned and, in many cases, the individual would not want to own them because it would not be economical to do so. In some situations it is more desirable to pay for their use.

TYPES OF LEASES

Leasing is an alternative to buying which should be considered where ownership is not a prerequisite for usage. The Financial Accounting Standards Board (FASB) in Statement No. 13 defines a lease as "an agreement conveying the right to use property, plant or equipment (land and/or depreciable assets) usually for a stated period of time."

The FASB classifies leases, from the standpoint of the lessee, as either capital leases or operating leases. A rule of thumb to distinguish a capital lease from an operating lease is: If the total lease payments are intended to enable the lessor to recover the original cost, it is a capital lease; if not, it is an operating lease. For accounting purposes FASB requires that any lease meeting its specific criteria for a capital lease must be capitalized on the balance sheet. In other words, the asset is treated much the same as an asset purchased by a debt issue as far as the balance sheet is concerned, and it is then depreciated in a manner consistent with the lessee's normal depreciation policy but with the lease term used for the amortization period. Take note, however, that for tax purposes it is the lessor who takes the depreciation (unless a special arrangement is made with the lessee).

An operating lease does not involve capitalization of the asset since neither ownership nor any arrangement equivalent to ownership is implied. Rental on an operating lease is charged to expense over the lease term as it becomes payable. The payments must be expensed on a straight-line basis unless it can be demonstrated that another basis is more consistent with use benefit from the leased property.

A consideration in whether to buy a property or to arrange for a capital lease would be the eligibility of the firm for the 10% investment tax credit. The credit is a reduction from federal income taxes in the year of acquisition, but unless the firm's profitability in that year is sufficient to enable it to use the credit there is no benefit. The credit cannot be applied to taxes in previous or succeeding years. If the firm cannot use the credit, therefore, a capital lease may be more economical. A lessor who could benefit from the credit could pass along some of the benefit to the lessee through advantageous lease terms.

LEVERAGED LEASES

Since our primary concern in this chapter is whether to lease or buy, the matter of how a lessor obtains financing is not relevant. However, since the leveraged lease has become so popular a brief explanation is provided.

A leveraged lease involves three parties—the lessee, the owner-lessor, and a third party who provides a substantial part of the financing by a loan to the owner-lessor. In financial terminology "leverage" refers to financing through debt rather than equity and the "leverage factor" is the debt/equity ratio. This type of lease is particularly beneficial to an owner-lessor since both the investment tax credit and the tax shelter from accelerated depreciation are available. Individuals in higher income tax brackets can thus obtain tax relief as owner-lessors in leveraged leases. Regardless of whether the lease is leveraged or not, the lease or buy analysis is not affected.

LEASE OR BUY

LCC analysis can be applied to the lease or buy decision with either the present worth or average annual cost method.

For the present worth method the buy option is evaluated by the usual technique and the lease option is assessed by obtaining the present worth of the lease payments. For the average annual cost method the UCR factor is applied to the present worth of each of the alternatives. If the lease payment is an annual charge, this figure is already an average annual cost.

EXAMPLE

The City of Harrisburg needs office space and has a choice of renting existing space at an annual charge of $40,000 or constructing a building at a cost of $300,000. It will need the space for 10 years, at which time the building will have a resale value of $180,000. Annual costs of the rented space will be $6,000, while similar costs of the new building will be $9,000. Financing will cost Harrisburg 10%. Compare the alternatives by the average annual cost method.

SOLUTION

Lease

Rent	$40,000	
Costs	6,000	
Annual Average Cost	$46,000	

Buy

P.W. of initial cost		$300,000
P.W. Salvage (180,000 × .38555)		(69,400)
		$230,600
230,600 × .16275	= $37,530	
+ annual costs	9,000	
Annual Average Cost	$46,530	

SUMMARY

Purchase	$46,530
Lease	46,000
Annual difference in favor of leasing	$ 530

DISCOUNT RATE

There is no universal consensus on what discount rate should be used for lease versus buy problems and the matter will not be debated here. Reliance is placed on the most commonly accepted method of determining the rate, that is, the after-tax cost of debt. For example, if the cost of debt financing for the project is 10% and the firm pays 50% income tax, the discount rate is 5%. If the firm's tax rate is 40%, the discount rate is 6%.

SELF-STUDY PROBLEMS

Problem 8-1.—The Upper Paxton Township Sewer Authority owns a system consisting of 100 miles of sewers. The system must be cleaned every 5 years and it is the practice of the Authority to clean 20 miles of it each year. A commercial contractor has done the work at a cost of $1,400 per mile. Would it be more economical to buy the equipment at a cost of $17,000? It is estimated that labor, maintenance, fuel, etc. would cost $20,000 a year. The equipment will last 10 years and have no salvage value. The appropriate discount rate is 10%.

Problem 8-2.—A borough manager was being urged by the road superintendent to purchase a new dump truck. He was reluctant to support the request because the truck would cost $25,000 and he figured that it would be used only about 30 days in a year. He could rent a similar truck for $100 a day.

(a) Make a list of all the pieces of information needed to solve this problem.

(b) Which of the two methods of LCC analysis is preferable for this problem?

Problem 8-3.—A manufacturer's representative must decide whether to buy a new automobile or lease one to conduct his daily business. He estimates he travels a total of 25,000 miles per year which includes approximately 25% personal usage. His tax bracket is in the low end of the 25% tax rate. If purchased, the car will be depreciated over a 3-year period and a straight-line method will be used. His before-tax interest rate is 12% per year.

He has obtained the following cost data to help him make the decision:

Lease Car

Mileage cost	$ .075/mile
Fixed charge	$200 per month

All other costs such as insurance and local taxes are paid by the leasing company.

Buy Car

Initial cost	$6,000
Operation and maintenance cost	$ .08 per mile
Annual insurance premium	$550
Annual local taxes	$24
Salvage value after 3 years	$2,000

Which is the better alternative?

Problem 8-4.—Blanchard Inc. needs office space and has a choice of renting existing space at an annual charge of $40,000 or construction of a building at a cost of $300,000. It will need the space for 10 years, at which time the building will have a resale value of $180,000. Annual costs of the rented space will be $6,000, while similar costs of the new building will be $9,000. A loan of $300,000 will cost Blanchard 10% interest. The company pays 40% tax on its income. It uses straight-line depreciation and will not be eligible for the investment tax credit on this purchase. Compare the alternatives.

Problem 8-5.—Thumann Co. is interested in a major piece of equipment which would cost $1,071,428. Its life is 5 years after which its salvage value will be $71,428. Maintenance costs will be $40,000 per year. Sum-of-the-years depreciation would be used. The firm's tax rate is 40%. Its borrowing cost is 10%. An investment tax credit of 10% is available. The alternative is a lease with an annual rental of $275,000 payable at the start of each of the 5 years. The lessor will pay the maintenance expenses. Compare the alternatives.

9

Replacements, Life Differences

REPLACEMENT

Replacement versus retention choices, in the broadest sense, include alternatives such as equipment versus equipment, equipment versus manual operation, equipment versus abandonment of the process, new equipment versus overhaul, etc. Regardless of the type of choice, however, the decision can be facilitated by life cycle cost analysis.

There are several approaches to replacement analysis but only one method is presented in this chapter—the method which the authors believe is the simplest and least prone to error. In all cases the replacement is treated as any new purchase. If the alternative is retention of old equipment, the latter is treated as a purchase also, its cost being the value that would be realized by its sale at the time of decision. Both the present worth and average annual cost methods may be used but the latter is generally more appropriate.

EXAMPLE

A motor is 11 years old and must now be rebuilt or replaced. Repairs will cost $100. If repaired, the life will be extended for 5 years. A new motor will cost $280 and will last for 12 years. Assume the standard salvage value of an old motor is $20. The discount rate is 10%. Should the motor be rebuilt or replaced?

SOLUTION

Average annual cost method:

New motor —

Initial cost	280 × (UCR, 12, 10%)	=	$41.09
Salvage	20 × (USF, 12, 10%)	=	.94
	AAC	=	$40.15

Old motor —

The value that can be realized on sale is $20.

Initial cost	20 × (UCR, 5, 10%)	=	$ 5.28
Repairs	100 × (UCR, 5, 10%)	=	26.38
Salvage	20 × (USF, 5, 10%)	=	3.28
	AAC	=	$28.38

It is more economical to rebuild the old motor.

A common feature of the replacement problem is that the projected lives of the alternatives are not the same. This difficulty was resolved in the example above by using the average annual cost method. Another technique is to compare the projects over only those years when the competition can exist, that is, the shorter of the two lives. An estimated salvage value is assigned to the longer-life asset at the end of the comparison period.

EXAMPLE

The Department of Transportation owns an earth mover which has an estimated remaining life of 5 years, after which it will be scrapped. A replacement would cost $60,000 and would have an expected life of 10 years after which it would be scrapped. The old machine can be sold now for $15,000. Annual costs for the new machine are estimated at $5,000 while the old machine will cost $14,000 a year. The estimated value of the new machine after 5 years is $15,000. The discount rate is 7%. Should the old machine be replaced?

SOLUTION

Old machine —

Initial cost $15,000 × (UCR, 5, 7%) = $ 3,658
Annual costs   14,000
                              AAC = $17,658

New machine —

Initial cost $60,000 × (UCR, 5, 7%) = $14,633
Annual costs   5,000
Salvage $15,000 × (USF, 5, 7%) =   2,608
                              AAC = $17,025

The average annual cost differential in favor of the new machine is $633.

TAX EFFECTS

It was stated that the value to be assigned to the old equipment as initial cost is the *value* that would be realized by its sale at the time of decision. For a tax-paying institution that value is not necessarily the same as the *price* that would be realized. Allowance must be made for tax effects of any gain or loss on the sale.

Gain or loss on a sale is the difference between the sale price and book value. Book value is the value of the asset on the company's accounting books and is the original cost less all the depreciation deducted since original purchase. An asset which originally cost $10,000 and is being depreciated by the straight-line method over 10 years has a book value of $6,000 after 4 years. If it can then be sold for $7,000 the difference of $1,000 is a gain and is subject to federal income tax at the company's tax rate. If it can be sold for only $5,000, the $1,000 loss can be deducted from income in calculating federal income tax.

The capital gains tax enters the picture only if the asset is sold for more than original cost. Then the ordinary tax rate would apply to the difference between book value and original cost and the capital gains rate would apply to the excess. In the above case, if the asset were sold for $11,000, the ordinary rate would apply to $4,000 and the capital gains rate to $1,000.

EXAMPLE

(a) Calculate the present worth of a compressor which has a book value of $500, can be sold now for $600, and which has a remaining life of 5 years. Allow for no salvage value. The firm's tax rate is 40% and its discount rate is 12%.

(b) Same problem but with a resale price of $400.

SOLUTION

(a) Initial cost is resale price less 40% of gain.
Initial cost $\quad\quad\quad\quad\quad$ $600 - $100 \times .40 \quad = $560
Depreciation tax benefit .40(100) (UPW, 5, 12%) = $\underline{144}$

$\quad\quad\quad\quad\quad\quad\quad\quad\quad\quad$ Present Worth = $416

(b) Initial cost is resale price plus 40% of loss.
Initial cost $\quad\quad\quad\quad\quad$ $400 + $100 \times .40 \quad = $440
Depreciation tax benefit .40(100) (UPW, 5, 12%) = $\underline{144}$

$\quad\quad\quad\quad\quad\quad\quad\quad\quad\quad$ Present Worth = $296

LIFE DIFFERENCES: CHAIN METHOD

An occasional difficulty encountered in comparing new equipment is that the expected lives are not identical. Under such circumstances the LCC analysis may be performed in the manner described above for replacements. An alternative approach, however, is the "chain method."

The chain method requires selection of a life cycle that is a common multiplier of the life of each alternative. For example, a life cycle of 30 years would be appropriate for alternatives with lives of 10 and 15 years. The 10-year-life alternative would be treated as three successive investments, all having the same costs. The 15-year-life equipment would be renewed once.

EXAMPLE

A state-owned health facility is faced with a decision on the purchase of a piece of equipment with an expected life of 6 years or one with an expected life of 4 years. The former (a) will cost $40,000 and will have annual operating and maintenance costs of $3,000 while the latter (b) will cost $30,000 and have annual operating and maintenance costs of $2,500.

Neither will have a salvage value. Compare by the present worth method using a 10% discount rate.

SOLUTION

The life cycle in this example is 12 years.

(a) Present Worth:

Initial cost		$40,000
Replacement	$40,000 × (SPW; 6, 10%) =	22,579
Operating and maintenance costs	3,000 × (UPW; 12, 10%) =	20,441
	Total =	$83,020

(b) Present Worth:

Initial cost		$30,000
Replacement 1	$30,000 × (SPW; 4, 10%) =	20,491
Replacement 2	30,000 × (SPW; 8, 10%) =	13,995
Operating and maintenance costs	2,500 × (UPW; 12, 10%) =	17,034
	Total =	$81,520

(b) is the more economical choice.

SELF-STUDY PROBLEMS

Problem 9-1.—A municipality's construction equipment manager must make a decision whether to replace a piece of equipment.

The following data has been obtained to help him make the correct decision:

Plan A — *Existing Equipment Cost Data*

Salvage value today	$10,000
Annual operating and maintenance cost	5,500
Expected additional life	5 years
Salvage value 5 years from today	3,000

Plan B — *New Equipment Cost Data*

Initial cost of equipment	$50,000
Trade-in value of existing equipment	15,000

Annual operating and maintenance cost $ 2,500/yr
Expected life of new equipment   25 years
Salvage value 25 years from today   3,000

Interest rate for both alternatives is to be 10%. Which is the better alternative based on the above data?

Problem 9-2.—(a) A school district is interested in determining the age at which it should replace its buses. As a start on the problem, use LCC to decide whether a typical bus should be replaced after 5 years, given the following information. The bus costs $20,000 and assume there is no change in price over the next 5 years. Also assume that a 5-year-old bus will be worth $10,000. Buses have a life span of 10 years. Annual costs for a bus (gas, maintenance, repairs, etc.) amount to $2,400 in the first year and increase by $180 a year thereafter. The district's discount rate is 8%.

(b) Using the same approach, how could one determine the optimum time for replacement?

Problem 9-3.—Alternative A has an initial cost of $15,000 and an annual operating cost of $1,500, an expected replacement cost of $17,000 and a salvage value of $5,000 at the end of 8 years. Interest is 8% per year.

Alternative B has an initial cost of $10,000 and an annual operating cost of $2,000, an expected replacement cost of $12,000 in 4 years and a salvage value of $3,000 at the end of each 4 year period.

SOLUTION

The life cycle in this example would be 8 years.

Alternative A

P.W. initial cost   = $15,000
P.W. operating cost = 1,500 × (UPW; 8, 8%)
                         = 1,500 × 5.74664   =   8,620
$17,000 need not be included in 8 year life cycle.
P.W. salvage   = 5,000 × (SPW; 8, 8%)
                        = 5,000 × .54027   = (2,701)
                Total P.W. of Alternative A = $20,919

Alternative B: To be completed by reader.

Problem 9-4.—Bloodhound Bus Corp. has the same problem as the school district in Problem 9-2. Its discount rate is 10% and it uses straight-line depreciation. Its income tax rate is 40%.

Problem 9-5.—*(a)* A polisher originally cost $9,000 and had an expected life of 15 years with no salvage value. Using straight-line depreciation, the firm now shows the book value at $6,000. Current market value of the polisher is $6,000. A new polisher is now available at a cost of $14,000 and it will have a life of 10 years with expected salvage value of $1,000. Its chief advantage is that it will provide pretax savings of $3,000 a year. If the federal income tax rate is 46%, the investment tax credit 10%, and the firm's discount rate 12%, determine if the replacement is advisable.

(b) Same problem except that the current market value of the polisher is $2,000.

10
Escalation

After a century of relatively stable prices, the United States, in the mid-'60s entered a period of pronounced inflation which has continued to the present and which shows no sign of terminating. Some of the story is portrayed in the charts in Figure 10-1. In "30 Years of U.S. Price History" the severity of the inflation is readily apparent in the average annual growths of 4.6% for 1967-72 and 7.7% for 1972-77. The growth rates of the Consumer Price Index on a yearly basis for 1970-78 are charted in "Another Outburst of Price Inflation." The trends in home prices are pictured in "Home Prices Keep Climbing" and a dismal outlook for the motorist is sketched in "Gasoline Prices Head Upward."

Indeed, inflation has become so much a part of the American way of life that planning which fails to allow for it is naive at best. There are sound life cycle costing procedures which can be utilized to minimize cost and which will blunt the impact of inflation. The purpose of this chapter is to explain the application of LCC to the purchase of assets which are affected by rising costs.

CAR PROBLEM

Let us return again to the car problem of Chapter 1, Example 1-1. On the basis of lowest initial cost, Car A proved to be the most economical. When subsequent costs were included with no allowance for the time value of money, Car C was superior. In Chapter 3 the discounting procedure was included and Car B had the least cost. In this chapter the final touch is provided by allowing for escalation of costs. Fuel costs

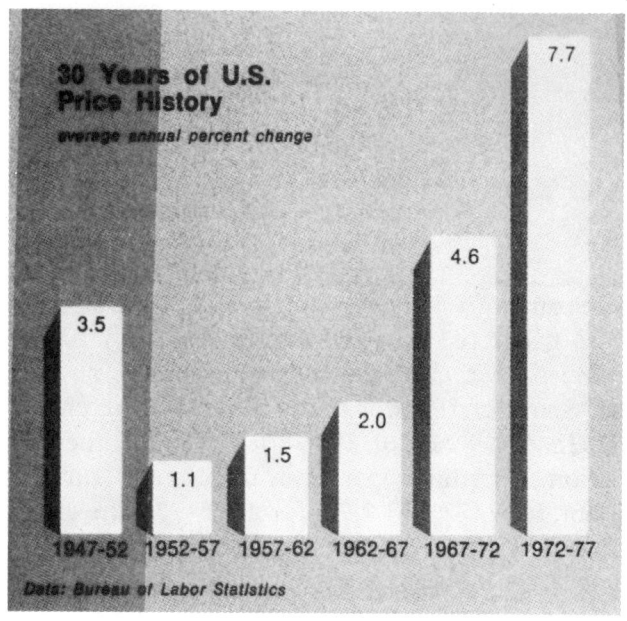
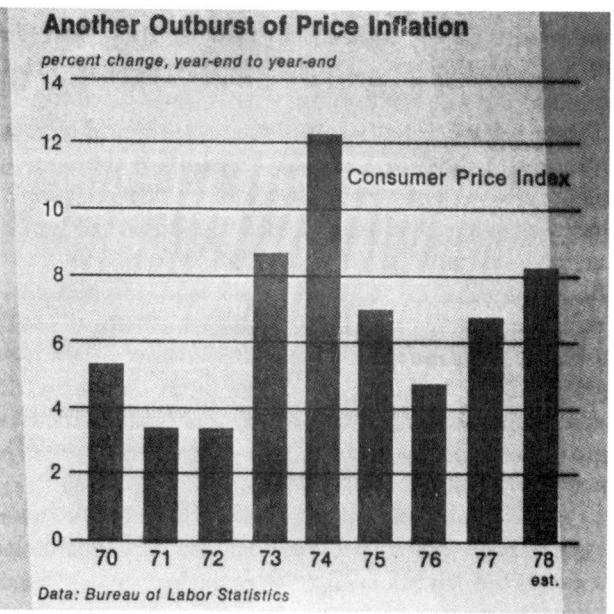

Figure 10-1. The Effects of Inflation

Escalation

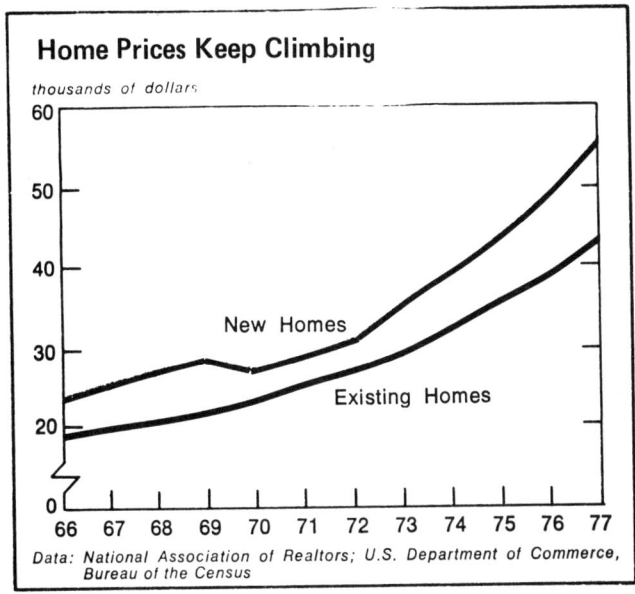

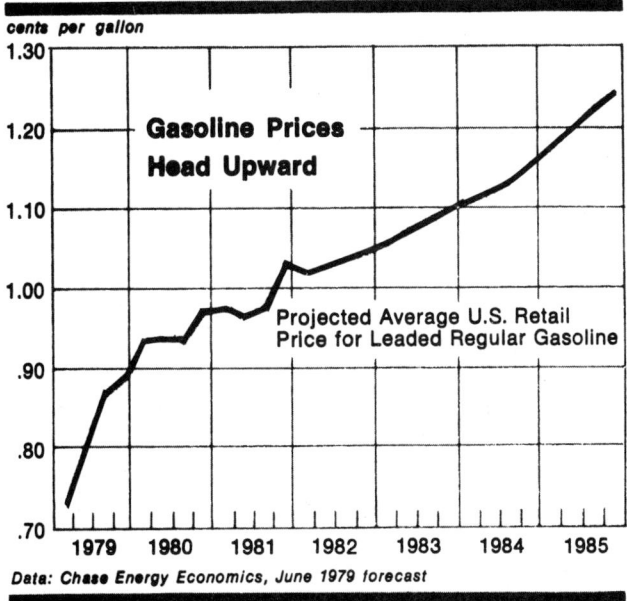

Figure 10-1 (continued)

From *Business Briefs,* courtesy of Chase Manhattan Bank, N.A.

122 Life Cycle Costing: A Practical Guide for Energy Managers

are assumed to escalate at an annual rate of 7% and maintenance and insurance costs at an annual rate of 5%.

SOLUTION

	CAR A	CAR B	CAR C
Purchase price	$3,200	$3,800	$3,910
Sales tax	160	190	195
Initial cost	$3,360	$3,990	$4,105
Present worth of Annual cost	CAR A	CAR B	CAR C
Fuel	$2,865	$2,150	$1,841
Maintenance	746	522	373
Insurance	712	824	1,220
License	79	79	79
PW of total annual cost	$4,402	$3,575	$3,513
Initial Cost	$3,360	$3,990	$4,105
	$7,762	$7,565	$7,618
Deduct Trade-In	(89)	(116)	(178)
Total LCC (PW)	$7,673	$7,449	$7,440
Differential	$ 273	$ 9	—

Best Choice: C, if only dollar costs are considered, A or B if the anticipated benefits are worth the differential. The supporting calculations are provided at this point, but it is suggested that the reader skip them until after reading the remainder of the chapter.

SUPPORTING CALCULATIONS

Car A

Fuel	733 × (DEF; 4, 8, 7) = 733 × 3.908284 = 2,865
Maintenance	200 × (DEF; 4, 8, 5) = 200 × 3.729833 = 746
License	24 × (DEF; 4, 8, 0) = 24 × 3.31212 = 79
Insurance	191 × (DEF; 4, 8, 5) = 191 × 3.729833 = 712

Trade-in $(100) \times \frac{(1.05)^4}{(1.08)^4}$ = (100) × .89 = (89)

Total PW of Annual Costs = 4,313

Car B

Fuel   550 × (DEF; 4, 8, 7) = 550 × 3.908284 = 2,150
Maintenance 140 × (DEF; 4, 8, 5) = 140 × 3.729833 = 522

| License | 24 × (DEF; 4, 8, 0) = | 24 × 3.31212 = | 79 |
| Insurance | 221 × (DEF; 4, 8, 5) = | 221 × 3.729833 = | 824 |

Trade-in $\quad (130) \times \dfrac{(1.05)^4}{(1.08)^4} \quad = (130) \times .89 \quad = \underline{(116)}$

Total PW of Annual Costs $\hfill = 3,459$

Car C

Fuel	471 × (DEF; 4, 8, 7) =	471 × 3.908284 =	1,841
Maintenance	100 × (DEF; 4, 8, 5) =	100 × 3.729833 =	373
License	24 × (DEF; 4, 8, 0) =	24 × 3.31212 =	79
Insurance	327 × (DEF; 4, 8, 5) =	327 × 3.729833 =	1,220

Trade-in $\quad (200) \times \dfrac{(1.05)^4}{(1.08)^4} \quad = (200) \times .89 \quad = \underline{(178)}$

Total PW of Annual Costs $\hfill = 3,335$

ESCALATING COSTS: PRESENT WORTH

Recurring costs which grow at a steady annual rate may be treated as an annually compounded series. If, for example, "e" represents the annual rate at which energy costs are expected to rise and "A" is today's cost of the energy needed annually by a project, then, assuming constant energy usage over the years, the costs may be represented as follows:

Year	Annual Cost
1	$A(1 + e)$
2	$A(1 + e)^2$
3	$A(1 + e)^3$
...	....
n	$A(1 + e)^n$

These annual costs are portrayed in Figure 10-2.

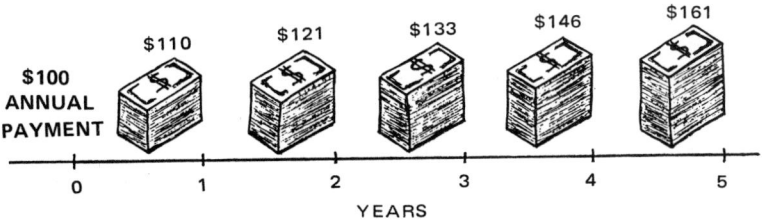

Figure 10-2. Escalation — 10% Per Year

The present value of these recurring costs will be:

$$P.W. = \frac{A(1+e)}{(1+i)} + \frac{A(1+e)^2}{(1+i)^2} + \ldots + \frac{A(1+e)^n}{(1+i)^n}$$

Formula 10-1

This present worth of escalating annual costs is portrayed in Figure 10-3.

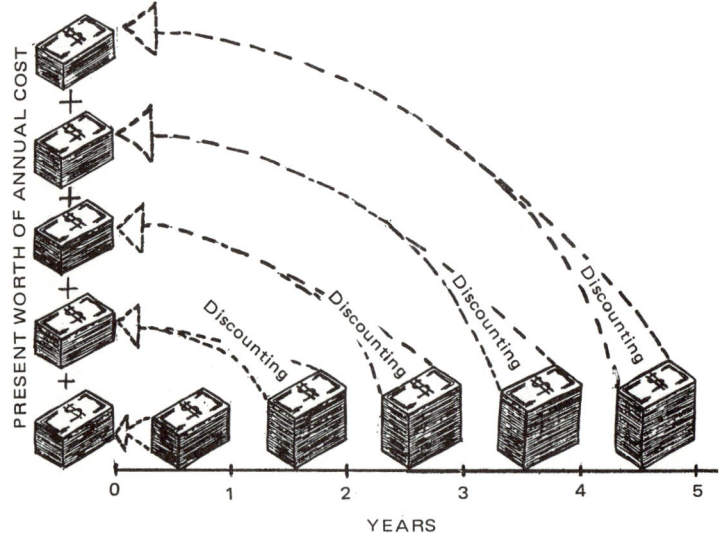

Figure 10-3. Escalating Annual Cost Discounted to Present Worth

EXAMPLE

Find the present value of the expected energy costs of a project having a planning horizon of 5 years if it will consume 500,000 KWH a year. Today's electricity price is $.04 per KWH, the expected escalation rate is 8%, and the discount rate is 10%.

SOLUTION

$$500,000 \times .04 = \$20,000$$

$$\underset{Year\ 1}{20,000 \frac{(1.08)}{(1.10)}} + \underset{Year\ 2}{20,000 \frac{(1.08)^2}{(1.10)^2}} + \ldots \underset{Year\ 5}{20,000 \frac{(1.08)^5}{(1.10)^5}}$$

or $20{,}000 \left[\dfrac{1.08}{1.10} + \dfrac{1.08^2}{1.10^2} + \dfrac{1.08^3}{1.10^3} + \dfrac{1.08^4}{1.10^4} + \dfrac{1.08^5}{1.10^5} \right]$

or 20,000 [4.35816] = $87,163

DISCOUNT-ESCALATION TABLES

The discount-escalation tables found in the Appendix provide an easy way to solve problems involving escalation. The annual escalation rates appear across the top of the page and the discount rates down the left margin. The numbers in the body of the tables are the present worth of a series of escalating payments compounded annually. The tables give values akin to UPW except that the series is escalating rather than uniform. Derivation of the formula for the tables may be found in Appendix A.

EXAMPLE

Use the discount-escalation tables to solve the preceding example.

SOLUTION

$20,000 × (DEF; 5 years, 10%, 8%) (found in Appendix)
20,000 × 4.35816 = $87,163

INTERVAL COSTS

Where the objective is to obtain the present worth of costs which are escalated at e% a year, compounded annually, but incurred at intervals other than one year, either of two methods may be used.

Method 1.—Consider each cost payment separately, escalating it to the year incurred and then discounting it to obtain the present worth.

For example: A cost of $70,000 is expected every 7 years in the life of an asset that will last for 20 years. The escalation rate is 8% and the discount rate is 14%.

(a) First payment —
$70,000 × 1.08^7 = amount of payment after 7 years
= 70,000 (SCA, 7, 8%) = $119,967

Now obtain the present worth of this amount.
$119,967 \times (SPW, 7, 14\%) = \$47,943$

(b) Second payment -
$\$70,000 \times (SCA, 14, 8\%) \times (SPW, 14, 14\%) = \$32,837$

Total of the two costs: $47,943 + 32,837 = \$80,780$.

Method 2.—If the costs are incurred every second or third year over a long life, Method 1 may become tedious. A shorter way of handling such problems is with the discount-escalation tables, but with a slight adjustment.

Let n = number of periods (i.e., payments)
Let p = number of years between payments
Adj. i is an adjusted i which may be used in the discount-escalation table.

$$\text{Adj. i} = \frac{(1+i)^p}{(1+e)^{p-1}} - 1 \qquad \text{Formula 10-2}$$

See Appendix A-2 for derivation.

EXAMPLE

Find the present worth of a cost of $1,000 occurring every 2 years for 20 years. Here, p = 2 and n = 10. The escalation rate is 9% and the discount rate is 10%. That is, i = .10 and e = .09.

$$\text{Adj. i} = \frac{(1.10)^2}{(1.09)^1} - 1 = .1100$$

Look in the n = 10 period, e = .09 column and the i = .11 row to find 9.060875 (table in Appendix).
Then, $\$1,000 \times 9.060875 = \$9,060.88$

EXAMPLE

Repeat the same example, but e = .06:

$$\text{Adj. i} = \frac{(1.10)^2}{(1.06)^1} - 1 = .14151$$

This requires interpolation as follows:

In the n = 10 period table in the e = .06 column, we find:

for i = .14		6.849390
for i = .15		6.564121
	Difference	0.285269
	X	.151
		0.043076
for i = .14151		6.849390
		−0.043076
		6.806314

Then, $1,000 × 6.806314 = $6,806.31.

ESCALATING COSTS: AVERAGE ANNUAL COST

Conversion of the present worth of escalating costs to an annual average cost is meaningful only if it is clear that the latter is a constant dollar figure expressed in the dollars of the time of purchase. This information may be useful for decision purposes but certainly not for any long-range budgetary planning.

SELF-STUDY PROBLEMS

Problem 10-1.—No. 2 fuel oil is currently selling at $.50 a gallon. A heating unit expected to use 40,000 gallons a year is being considered. What is the present value of the oil to be used over its 15-year planning horizon if oil prices are expected to increase at 5% a year and the buyer's discount rate is 10%?

Problem 10-2.—A large municipality is considering replacing 288 incandescent lighting fixtures in the automotive fleet's garage. The light bulbs are 150 watts rated at 2,100 lumen/lamp output and are energized 24 hrs/day, 365 days/year. Cost of electricity to the state is $.03/KWH. Lamp life is 2,500 hours and the lamps cost $1.18 each.

The alternatives to the incandescent lamps would be:

(a) Circleline fluorescent lamps that can be screwed into the same lighting fixture, rated at 60 watts per fixture, costing $34.85 each installed, life of 22,500 hours and rated at 2,550 lumens/lamp output. Cost for replacement lamp/fixture is $13.15.

(b) 40-watt, 4-foot strip lighting fixture. The incandescent fixtures must be replaced with 4-foot strip fixtures at an initial cost of $25.00 each installed. Life of 36,000 hours and 2,770 lumens/lamp output. Cost for 1 lamp fixture is $1.53.

Assume:

Since lumen output is approximately equal for all three lamps, assume all lamp output to be equal. Relamping labor cost for all three fixtures is $.50/lamp. Electricity will escalate at 7% per year and labor and lamps cost will escalate at 5% per year. Use a 7%/yr discount rate.

Problem 10-3. Air Pollution—A power company recently was faced with the decision on whether to purchase an electrostatic precipitator or a fabric filter bag collector in order to satisfy exhaust control requirements. The company was using lignite coal with a .67% sulfur in a 525 MW boiler and both types of equipment could handle the task.

Specifications on each as follows:

EP:	Total initial costs*	$14,100,000
	Operating costs*	
	Energy 4876 KWH @ 2.682¢	1,145,583/yr
	4876 × 365 × 24 hr × 2.682	
	Surveillance – 1 hr per day	5,110/yr
	Maintenance costs*	
	Defective wires, repairs	68,000/yr
	Cleaning	800/yr
	Repairs	480,000/7 yrs
	Miscellaneous maintenance	5,000/yr
FFBC:	Total initial costs*	12,500,000
	Operating costs*	
	Energy 3376 KWH @ 2.682¢	793,168/yr
	Surveillance – 1 hr per day	5,110/yr
	Maintenance costs*	
	Bag replacement	636,768/2 yrs
	Repairs, cleaning, etc.	50,000/yr
	Repairs	70,000/7 yrs

*20-year life, 8% inflation rate, 14% cost of capital

(a) Solve without tax considerations.

(b) Solve allowing for straight-line depreciation and a 40% tax rate. The investment tax credit is 10%.

Problem 10-4. Telephone Installation—Two different types of a telephone system are under consideration for installation in a one-story office building. The gross floor area measures 90 feet by 140 feet. There is a central service core area that is 30 feet by 60 feet. For both installations, two telephone closets will be needed at both ends of the service core with distribution wiring to the office areas originating at these closets.

Historically, one telephone will be installed for about every 100 square feet of usable floor space. With a net floor of 10,800 square feet, the ultimate number of telephones for this office building will be 108.

The average office telephone is relocated every two years (about one-half of the telephones are relocated each year, or 54 telephones per year).

The specifications for the two systems are as follows:

(a) *Underfloor Duct System*—Rectangular ducts with preset inserts (access collars) in parallel rows six feet apart. These ducts are connected to each other and to the telephone closet by header ducts.

Unit Cost/gross sq ft	$ 1.00
Relocation service charge/telephone	52.50

(b) *Ceiling Zone System*—Metal conduits run above suspended ceiling from the closets to the center of a telephone zone (area 400–600 sq ft). Pathway is then continued from the zone center to each desk location via branch conduit and a telephone (and power) service pole.

Unit Cost/gross sq ft	$ .68
Relocation service charge/telephone	92.50

Using a 20-year life, a discount rate of 8%/yr and an assumed 10%/yr escalation rate for the service charges, compare the two installations.

11

Payback, Break-Even Analysis

PAYBACK

The payback period is the length of time necessary to recover the initial investment of a project. The funds available for the recovery are the total of net savings on an after-tax basis and the depreciation tax benefit. Financing costs are not included.

Simple Payback

The simplest form of payback makes no allowance for the time value of money. The only concern is: How fast can individuals get back their investment? If an investment of $15,000 provides after-tax savings of $5,000 a year, the simple payback is 15,000 ÷ 5,000 = 3 years.

EXAMPLE

Find the simple payback for an outlay of $10,080 for equipment having a life of 7 years. Savings before taxes will be $4,000 annually, the firm's tax rate is 40%, and the sum-of-years digits depreciation is used.

SOLUTION

Year	Depreciation Tax Benefit	Savings After Taxes	Total	Cumulative
1	7/28 X 10,080 X .4 = 1008	2,400	3,408	3,408
2	6/28 X 10,080 X .4 = 864	2,400	3,264	6,672
3	5/28 X 10,080 X .4 = 720	2,400	3,120	9,792
4	4/28 X 10,080 X .4 = 576	2,400	2,976	

Answer: $3 + \dfrac{288}{2976} = 3.1$ years

In this problem the annual savings were constant throughout the period. Had inflation been considered it is possible that a trend in annual savings would have existed. If the above equipment could have produced progressively larger savings from cost differentials the payback would have been smaller. Let's say that annual savings before taxes were expected to increase at 6% a year. The results:

Year	Depreciation Tax Benefit	Savings After Taxes	Total	Cumulative
1	1,008	2,400	3,408	3,408
2	864	2,544	3,408	6,816
3	720	2,696	3,416	

$10{,}080 - 6{,}816 = 3{,}264$

Answer: $2 + \dfrac{3{,}264}{3{,}416} = 2.96$ years

The cumulative savings over the 7 years, without escalation, amount to $20,832, or more than double the initial investment. However, the firm which uses a 3-year cutoff on payback would reject the project and thus fail to achieve this benefit. The lack of concern for returns after the payback period is a major drawback of the simple payback criterion.

Discounted Payback

A second criticism of the simple payback is that it ignores the time value of money. This defect, however, can be overcome by using the discounted payback. The discounted payback differs from simple payback in that the returns are discounted.

EXAMPLE

Find the discounted payback for an outlay of $10,000 for equipment having a life of 8 years. It will produce annual savings of $3,000 after taxes and including the depreciation tax benefit. The discount rate is 10%.

SOLUTION

Year	Discounted Savings	Cumulative Discounted Savings
1	$\frac{3{,}000}{1.10} = 2{,}727.27$	2,727.27
2	$\frac{3{,}000}{1.10^2} = 2{,}479.34$	5,206.61
3	$\frac{3{,}000}{1.10^3} = 2{,}253.94$	7,460.55
4	$\frac{3{,}000}{1.10^4} = 2{,}049.04$	9,509.59
5	$\frac{3{,}000}{1.10^5} = 1{,}862.76$	

$10{,}000 - 9{,}509.59 = 490.41$

Answer: $4 + \frac{490.41}{1862.76} = 4.26$ years

LOGARITHM METHOD

The above method for calculating discounted payback can become tedious for projects having long lives. A simpler approach is the logarithm method. But don't flip the page if you forgot your logarithms because you don't really need them. Just refer to Figure 11-1, a Payback Period chart.

The derivation of this logarithm formula is in Appendix A.

Let n = number of years for payback
s = annual savings
c = cost
i = discount rate

$$n = \frac{\log \frac{1}{1 - \frac{ci}{s}}}{\log (1 + i)}$$

Formula 11-1

e.g., let c = $10,000; s = $3,000; i = .10

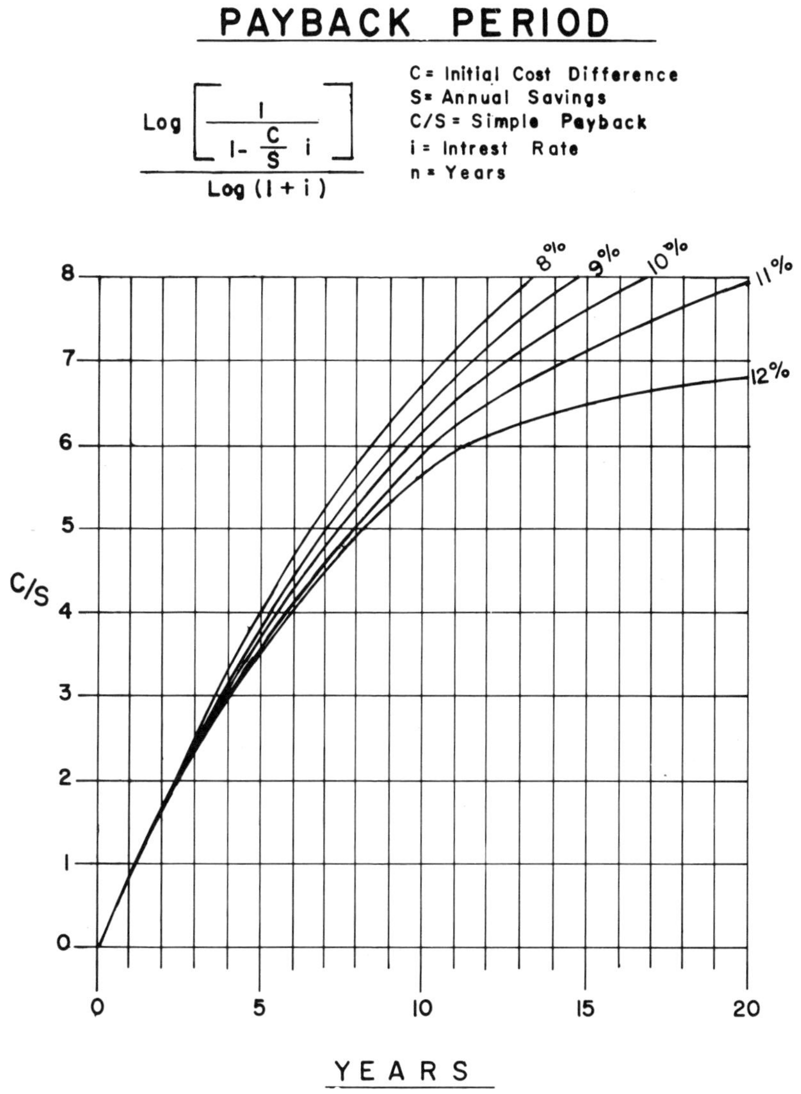

PAYBACK PERIOD
Log [1 / (1- C/S i)] / Log (1 + i)
C = Initial Cost Difference
S = Annual Savings
C/S = Simple Payback
i = Intrest Rate
n = Years
8%
9%
10%
11%
12%
C/S
0
1
2
3
4
5
6
7
8
0
5
10
15
20
YEARS

Figure 11-1.

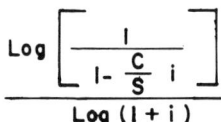

$$n = \frac{\log \dfrac{1}{1 - \dfrac{10(.10)}{3}}}{\log 1.1} = \frac{\log 1.5}{\log 1.1} = 4.25$$

Logarithm tables are not included here. Use a log-capable calculator or an auxiliary set of log tables. The Payback Period chart in Figure 11-1 is based on the logarithm formula. Take a look at it.

The left margin is c/s. The c/s is merely the simple payback calculated by dividing the annual savings into the cost. The curved lines represent the discount rates. For this problem c/s = 10/3 = 3-1/3 and i = 10%. With a ruler pinpoint the spot on the 10% curve that lines up with c/s = 3-1/3. Immediately below this spot on the horizontal axis is the payback—in this case 4.25 years. That's all there is to it.

TRUE PAYBACK

True payback is discounted payback expanded to allow for escalation. Here, too, the formula is based on logarithms, but the reader may skip down to the paragraph explaining the True Payback chart (Figure 11-2).

In the following formula c/s is the simple payback period and $k = \dfrac{1 + e}{1 + i}$ where e is the escalation rate and i is the discount rate.

$$n = \frac{\log 1 + \dfrac{c}{s}\left(1 - \dfrac{1}{k}\right)}{\log k} \qquad \textit{Formula 11-2}$$

EXAMPLE

The simple payback for a project is 4 years, the escalation rate for savings is 2%, and the discount rate is 10%. Find the True Payback.

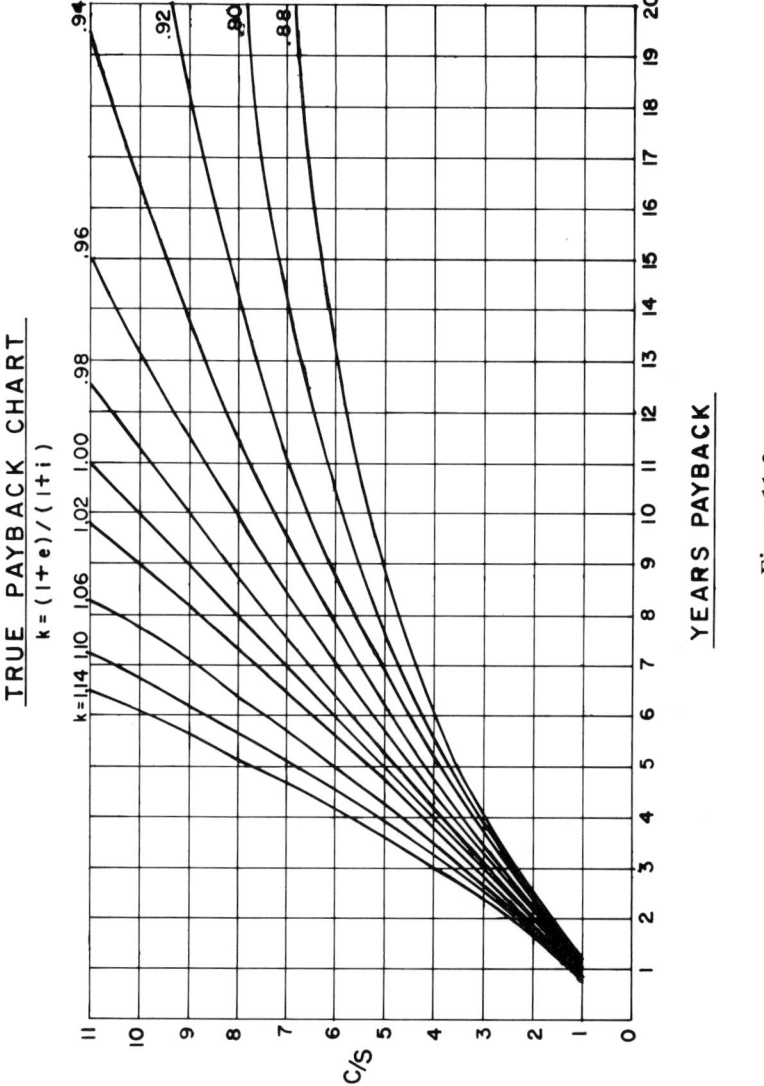

Figure 11-2

SOLUTION

$$\frac{c}{s} = 4 \text{ and } k = \frac{1.02}{1.10} = 9.27273 \text{ and } \frac{1}{k} = 1.07843$$

$$n = \frac{\log[1 + 4(1 - 1.07843)]}{\log .927273} = 4.986 \text{ or about 5 years.}$$

Let's solve this problem with the True Payback chart (Figure 11-2). The left margin is the simple payback, that is, $c/s = 4$. The curved lines represent k and k is $\frac{1+e}{1+i} = \frac{1+.02}{1+.10} = .927$. Locate the intersection of $c/s = 4$ and $k = .927$ and find the True Payback immediately below on the horizontal axis. The answer is 5.0.

The True Payback chart can be used in place of the Payback Period chart since it covers cases where the escalation rate is zero.

BREAK-EVEN ANALYSIS

In the section just completed the type of question asked was: "Using the discounted payback method find the number of years required to recover the following invest . . . ." Had the question been worded: "Find the break-even number of years for the following investment . . . ." the solution would have been exactly the same but we would have called the process "break-even analysis." In other words, discounted payback is break-even analysis.

Break-even analysis, however, is of broader scope than payback since its objective is to find the break-even point in terms of any variables or parameters that the analyst may have in mind. It deals with such questions as:

1. Given such-and-such information, in how many *years* will the project break even?
2. Given the following, at what *discount rate* will the project break even?
3. Given thus-and-such, how many *miles* of driving will be required to break even?
4. Given these circumstances, how many *people* need to be processed to break even?

EXAMPLE

A hospital is considering purchase of equipment costing $100,000. Treatment will be given to 2,000 people per year and the equipment should be useful for 20 years. Associated fixed costs will be $21,000 per year and the variable costs will be $2.00 per treatment. What charge per treatment will enable the hospital to break even? The discount rate is 10%.

SOLUTION

(a) By the average annual cost method:

Convert the initial cost to its equivalent average annual cost and add the annual fixed and variable costs—

Average annual cost = 100,000 × (UCR, 20, 10%) + 21,000 + $2 (2,000) = $36,746.

See Figure 11-3.

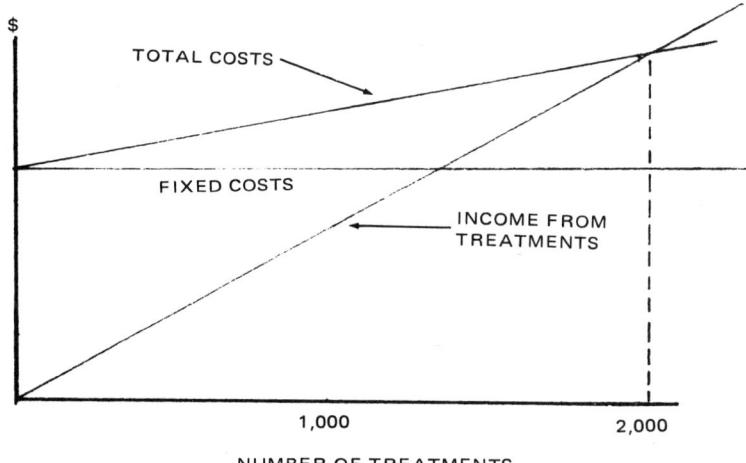

Figure 11-3

Now divide the average annual cost by the number of treatments to obtain the break-even charge:

$36,746 ÷ 2,000 = $18.38

(b) By the present worth method:

Obtain the present worth of total costs by adding initial cost to present worth of the annual costs—

P.W. of total costs = $100,000 + [25,000 × (UPW, 20, 10%)]
= $312,839

Next, let C represent the total annual charge and find the present worth of the income—

P.W. of income = C × (UPW, 20, 10%)
= C × 8.51355

Then, P.W. of income = P.W. of total costs

C × 8.51355 = 312,839
and C = $36,746

$$\text{Charge per treatment} = \frac{C}{2,000} = \frac{36,746}{2,000} = \$18.38.$$

EXAMPLE

A township is considering purchase of a truck for $10,000. It is estimated that the truck will be retained for 5 years and then sold for $5,000. Savings over a rental agreement will be $4,500 a year. The township authorities do not know what discount rate to use but believe they can make a decision if they know the rate at which the buy versus rental alternative will break even. Find the rate.

SOLUTION

Buy Alternative		Lease Alternative
$20,000 − 5,000 (SPW, 5, i%)	=	4,500 (UPW, 5, i%)

where i represents the rate at the break-even point.

The following is obtained by trying various rates in the above equation:

i%	Costs	Savings	Difference
9%	16,750	17,503	753 excess savings
10%	16,895	17,058	163 excess savings
11%	17,032	16,631	401 excess costs

Slightly above 10%. A more accurate answer is 10.3% which can be obtained with the same technique between 10% and 11% using the formulas for SPW and UPW in Chapter 2.

SELF-STUDY PROBLEMS

Problem 11-1.—Determine the simple payback of an energy-saving project requiring an initial outlay of $80,000 and expected to last for 8 years, at which time the salvage value will be zero. It can produce net annual pretax savings of $20,000. The firm's income tax rate is 40% and it uses double declining balance depreciation. Its cost of capital is 10%.

Problem 11-2.—Determine the discounted payback of Problem 11-1.

Problem 11-3.—Using the logarithm method, determine the payback for a project having a cost of $38,000 and total annual recovery of $12,000 over the next 8 years. The firm's discount rate is 10%. Check answer on Payback Period chart (Figure 11-1).

Problem 11-4.—An engineering consulting report has defined payback as the time to reach the point at which the accumulated savings equals the original investment invested at the interest rate. Is this a correct statement?

Problem 11-5.—A 60,000 sq ft state office building is currently in the design stage, and under consideration for the lighting system is a ballast which will cut energy consumption by .15 watts per sq ft. Today's cost of electricity is $.03 per KWH, the expected escalation rate is 7%, and total hours of operation per year will be 3200. If the system is installed, the savings on air conditioning will be one-fourth of the KWH saved on lighting. The additional cost of the system is $6,000. What is the discounted payback using a 9% discount rate?

Problem 11-6.—A homeowner, patriotically, as well as economically, motivated, is contemplating installation of a solar system in a house he is building. The house will require 400 sq ft of collector area, the cost of which will be $20 per sq ft of collector area. The cost will be included in the 30-year mortgage at 9%. Annual savings on energy is expected to be $1.20 per sq ft of collector area. The expected inflation rate on energy costs for the foreseeable future is 6%. A HUD grant of $400 and a 30% energy tax credit will be available. Should the owner buy the system if he requires a discounted payback of 15 years?

140   Life Cycle Costing: A Practical Guide for Energy Managers

Problem 11-7.—Two different 25-HP motors are under consideration for a boiler plant of a central steam distribution system. One electric motor manufacturer certifies that his energy efficient motor has an efficiency of .914 and cost $739.00. A different manufacturer certifies that his standard motor's efficiency is .882 but will cost only $628.00.

If energy cost is $.04/KWH and the motor operates at full load, what is the break-even number of hours per year of operation? Life expectancy of both motors is 30 years and their maintenance is expected to be the same. The appropriate discount rate is 10%. (1 HP = .746 KW.)

Problem 11-8.—An immunization plan is being considered by a community health center. Its initial cost will be $40,000 and annual costs thereafter will be $14,000 until the disease is eradicated. Savings to the center for treatment of patients will be $14,000 a year. In how many years will the center break even? Its discount rate is 10%.

Problem 11-9.—The owner of a small business is deciding on whether to buy or rent a delivery truck. If purchased, the truck will cost $12,000 and will be depreciated over 5 years by straight line. The fixed charge for leasing will exceed fixed costs of ownership by $4,400 per year. Mileage cost for leasing is $.14 per mile while operating and maintenance cost of ownership is $.16 per mile. Salvage value of the truck after 5 years will be $2,000. The owner's effective tax rate is 50% and his discount rate is 12%. Determine the break-even mileage per year assuming that the truck is driven the same number of miles each year.

12

Ranking

Earlier in this text a problem involving a choice of car was resolved by using the life cycle costing method of ranking. The fact that the cars were mutually exclusive (that is, if one car were chosen the others would not be) made the process of selection a simple one. Assuming an unconstrained budget (the ability to pay the initial cost of any of the cars) eliminated another factor which could have affected the process. But alternatives are not always mutually exclusive and budgets are often inhibiting and ranking can become more complex than it was with the car problem.

NET PRESENT WORTH (VALUE) METHOD

Ranking problems usually arise in one of two situations—purchases of new assets or retrofits. In both cases the savings must justify the marginal outlay, but the choice among alternatives is not always obvious. An example may make the point.

A choice is to be made among four cars, three new cars and the fourth a used car which originally cost $30,000. The buyer expects to drive the chosen car 10,000 miles a year for 5 years and then dispose of the car. The buyer's cost of money is 10%.

Information	A	B	C	D
Initial Cost	$ 5,000	$ 5,300	$ 5,200	$ 8,000
Average Annual Costs	1,800	1,600	1,660	1,800
Disposal Value	0	0	0	6,200

Using the net present worth method of ranking, car D is favored:

Computations	A	B	C	D
Initial Cost	$ 5,000	$ 5,300	$ 5,200	$ 8,000
P.W. of Average Annual Costs	6,823	6.065	6,293	6,823
P.W. of Disposal Value	0	0	0	(3,850)
Net P.W. of Costs	$11,823	$11,365	$11,493	$10,973
Ranking:	4	2	3	1

A budget constraint of $6,000 would, of course, eliminate car D from consideration in the first place. Assuming no such constraint, however, the buyer may still be reluctant to buy car D because the payback period is too long. Payback will take 5 years since the $3,000 marginal payment ($8,000 for car D less $5,000 for car A, the lowest-priced car) will not be recovered until the time of disposal.

Car B ranks second by the NPW method, but its payback is only 1½ years ($300 marginal cost difference paid back by $200 average annual cost savings). A combination of NPW and payback methods will, therefore, establish car B as the best choice of A, B, and D. Car C, however, although ranked third by NPW, has the best payback of all at 1.43. The question arises: Should payback supercede NPW as a ranking criterion? If so, car C should be bought.

Payback is not a desirable ranking method because it ignores savings expected after the payback period. A payback period limitation may be set to correspond with the time limit of the loan or to satisfy some other purpose, but such a constraint will do no more than confine the selection to the eligible choices. In the car problem, if the payback limit had been 3 years, the choice would have been limited to cars B and C, and car B would have been chosen on the basis of the NPW ranking.

NPW ranking is useful but it is not always satisfactory. The situation could exist where two projects satisfied the payback limitation and provided equal net present value of savings and yet the cost of one project considerably exceeded that of the other. The ranking method which is sensitive to such problems is the savings/investment ratio.

SAVINGS/INVESTMENT RATIO METHOD

The savings/investment ratio (SIR) is the ratio of the net

present worth of savings to the net present worth of investment costs. The denominator includes initial cost, present worth of subsequent investment and replacement costs, and present worth of salvage value (as a deduction). The numerator is the present worth of the annual cost savings. The ratio obtained is also known as the benefit/cost ratio. Projects are ranked from highest to lowest SIR and all those with an SIR less than 1 are rejected.

EXAMPLE

A firm located in Minnesota installed a shelter on its loading dock to reduce heat loss. The cost was $1,500. Calculations based on dock area, air flow, and temperatures revealed an energy loss of 300 million Btus per year. A 70% efficiency in energy saving from the shelter was assumed. Calculate the SIR on the basis of 8 years, a discount rate of 10%, and a cost of $.65 per gallon of No. 2 oil.

SOLUTION

Energy saved annually: $\dfrac{300{,}000{,}000 \times .70}{138{,}000} = 1{,}522$ gals. oil

(1 gal. No. 2 oil = 138,000 Btus)
Cost savings: $1{,}522 \times .65 = \$989$

$$\text{SIR} = \dfrac{989 \times (\text{UPW}, 8, 10\%)}{1{,}500} = 3.5$$

SIR — Car Problem

Returning to the car problem, the SIRs are as follows:

Car B: $\dfrac{(6823 - 6065) \times (\text{UPW}, 5, 10\%)}{300} = 9.58$

Car C: $\dfrac{(6823 - 6293) \times (\text{UPW}, 5, 10\%)}{200} = 10.04$

Car D: $\dfrac{0}{-850} = \ ?$

The difficulty with car D is that the savings are realized in the resale value and do not enter the numerator. This is a weakness of SIR which can only be circumvented in cases such as this by recognizing that resale is a source of savings and including its present worth in the numerator rather than the denominator. The SIR for car D would then become 3850/3000 = 1.28. The final ranking by SIR would give car C a slight edge over car B and both would be considerably ahead of car D.

BTU METHOD

The annual Btus saved per investment dollar is a ranking criterion that may be useful where energy conservation is of prime concern. Calculating this savings can be illustrated by the loading dock problem cited above:

$$\text{Btus per investment dollar} = \frac{300{,}000{,}000 \times .70}{1{,}500} = 140{,}000$$

This method has several weaknesses:

1. It makes no allowance for the costs of different types of energy;
2. It does not take into account nonenergy savings and costs;
3. It ignores expected life of the project and the time value of money.

Although this technique may yield the largest annual Btu energy savings per dollar invested, it will not necessarily provide the largest dollar or energy savings for a given budget.

REQUIRED RANKING

The Federal Energy Management Program (FEMP) requires that the savings/investment ratio be used to rank retrofit projects to determine funding priorities. Where projects have identical SIRs, the Btu and discounted payback measures are recommended as secondary criteria. For new buildings, FEMP requires that priority be based upon lowest total life cycle cost.

MULTIPLE PROJECTS

Up to now in this section, the objective of ranking has been to select the most desirable of several projects which were mutually exclusive. Where more than one project can be ac-

Ranking

cepted, problems of interdependence and budget constraints may make the ranking process more complex. Standardized decision rules may not always produce optimal results.

EXAMPLE

An owner of a building is contemplating the installation of two energy saving projects. He will use the SIR criterion to determine if the projects are cost-effective. If both projects prove to be cost-effective, both will be implemented. The projects are:

1. Increase the insulation of the building envelope which would mean a savings of 93,340,000 Btu/yr. The cost to install the insulation would be $7100.
2. Install a new burner which would improve the burning efficiency from 60% to 70%. Cost to install a new burner would be $2500. The building's existing heat loss is 250,000,000 Btu/yr.

The owner realizes that these two projects are interdependent and he will base his decision on which project to accomplish first, using the savings to investment ratio criterion.

The evaluation period will be 20 years, the discount rate is 10%, and No. 2 oil with 138,000 Btu/gal. is used to heat the building. Oil cost is $.60/gal. and the escalation rate for oil is estimated at 7%/yr.

Which project(s) should be implemented?

SOLUTION

Evaluating each project as an independent project, the SIR of each is determined as follows:

1. *Install Insulation*

$$\frac{\$ \text{ saved}}{\text{yr}} = \frac{Q \times \text{Unit Price}}{\text{Fuel Btu per unit} \times \text{Efficiency of burner}}$$

$$\frac{\$ \text{ saved}}{\text{yr}} = \frac{93{,}340{,}000 \times .60}{138{,}000 \times .60}$$

$$\frac{\$ \text{ saved}}{\text{yr}} = \$676$$

$$\text{SIR} = \frac{\$676 \times (\text{DEF}; 20, 10, 7)}{\$7100}$$

$$\text{SIR} = \frac{\$676 \times 15.15114}{\$7100} = 1.44$$

2. *Installing a More Efficient Burner*

$$\frac{\$ \text{ saved}}{\text{yr}} = \frac{250{,}000{,}000 \times .60}{138{,}000} \left[\frac{1}{.60} - \frac{1}{.70} \right]$$

$$\frac{\$ \text{ saved}}{\text{yr}} = \$259$$

$$\text{SIR} = \frac{\$259 \times 15.15114}{\$2500} = 1.57$$

When analyzed as independent projects, both installations are cost-effective and should be implemented based on the SIR analysis technique. Since the project to improve the burner efficiency has a higher SIR, a logical decision would be to implement this project first.

If the burner project were first implemented to increase its efficiency to 70%, the SIR for installing insulation would change as follows:

$$\frac{\$ \text{ saved}}{\text{yr}} = \frac{93{,}340{,}000 \times \$.60}{138{,}000 \times \$.70}$$

$$\frac{\$ \text{ saved}}{\text{yr}} = \$580$$

$$\text{SIR} = \frac{\$580 \times (\text{DEF}; 20; 10, 7)}{\$7100}$$

$$\frac{\$580 \times 15.15114}{\$7100}$$

$$= 1.23$$

Insulation is still cost-effective.

Now an analysis of installing a more efficient burner when the insulation is installed should be made.

$$\frac{\$ \text{ saved}}{\text{yr}} = \frac{(250{,}000{,}000 - 93{,}340{,}000) \times .60}{138{,}000} \left[\frac{1}{.60} - \frac{1}{.70} \right]$$

$$\frac{\$ \text{ saved}}{\text{yr}} = \$162$$

$$\text{SIR} = \frac{\$162 \times 15.15114}{\$2500}$$

$$\text{SIR} = .98$$

CONCLUSION

Although the installation of a more efficient burner has a higher SIR independently, it is not cost-effective if the insulation is installed.

ESCALATION

Escalation of costs and savings have an impact on SIR. As an illustration, let us again return to the problem analyzed earlier in this chapter on the loading dock. The cost (denominator) was $1,500 and the annual savings without escalation (numerator) were $989. The SIR was 3.5. Now let us assume an annual oil price escalation of 6%.

$$\text{SIR} = \frac{989 \times (\text{DEF}; 8, 10, 6)}{1,500} = \frac{989 \times 6.7963}{1,500} = 4.48$$

Inclusion of the escalation factor improved the SIR considerably.

DEPARTMENT OF ENERGY METHOD

The U.S. Department of Energy has developed projections of energy prices for the various types of fuels for the residential, commercial and industrial sectors and by geographical regions. On the basis of these projections it has produced sets of modified uniform present worth tables which are similar to our discount-escalation tables, but which have the appropriate energy escalation rates incorporated in them. The user merely needs to find the particular modified uniform present worth factor (abbreviated UPW*) in a table and multiply by the base year cost. The 10% discount rate prescribed by the Office of Management and Budget (OMB) is used for all of the tables. These tables are to be used in evaluation of projects in Federal buildings.

A basic difference between our treatment of escalation and that of the Department of Energy should be clarified. The discount rate of 10% required by OMB is an after-inflation rate and ours is not. Their energy escalation rate is the rate above the general inflation rate and ours is not. To illustrate: Let's say the general inflation rate is 5% and the energy escalation rate is 9%. Since opportunity rates in the open market always include an allowance for inflation, our method would use 15% as the discount rate. We would use 9% as the energy escalation rate. The Department of Energy uses 10% as the discount rate and a 4% energy escalation rate.

EXAMPLE

Installation of a central automatic environmental control system in a building is expected to cost $5,000. It is expected to last 20 years and to have a salvage value of $2,000 at that time. It will increase maintenance costs by $200 per year and will require an outlay of $800 for parts replacement at the end of year 10. It will reduce annual energy usage by 200M Btus. The building is located in DOE Region 2* and is heated and cooled by distillate fuel oil purchased at the industrial rate of $4.44. Discount rate is 10%. Calculate the SIR of this project.

SOLUTION

Numerator

Energy savings:	200M Btus × $4.44 × 11.056	= $9,818
Maintenance costs:	$200 × (UPW, 20, 10%)	
	200 × 8.51355	= 1,703
	Annual savings	= $8,115

Denominator

Initial cost		= $5,000
Salvage:	2,000 × (SPW, 20, 10%)	
	2,000 × .14865	= 297

The UPW for Region 2, industrial sector, distillate oil, 20 years, is 11.056. This value and the $4.44 price were obtained from "Federal Energy Management and Planning Programs; Proposed Methodology and Procedures for Life Cycle Cost Analysis of Federal Buildings," published in the *Federal Register*, April 30, 1979.

Ranking   149

Replacement: 800 × (SPW, 10, 10%)
                    800 × .38555   =   308

                                             $5,011

        SIR = 8,115 ÷ 5,011 = 1.62

SELF-STUDY PROBLEMS

Problem 12-1.—Consider the following investment opportunity:

Initial cost	= $150,000
Additional cost at the end of year 1	= 50,000
Savings at end of year 1	= 0
Annual savings at end of year 2–10	= 40,000
Salvage value of investment	= 0

With a discount rate of 10%, what is the savings to investment ratio if no escalation is considered?

Problem 12-2.—Using an annual escalation rate of 8% applied to the savings and 5% escalation rate applied to the additional cost at the end of year 1, determine the SIR for Problem 12-1.

Problem 12-3.—National Corporation is considering two energy conservation–oriented projects for implementation. Due to budgetary restraints, only one project of the two can be funded and they are not interdependent in their savings.

The first project is to insulate a steam line at a cost of $20,000 and will realize an annual after-tax savings of $6,000 in energy costs.

The second project is to install a wind shelter to a loading dock area which will cost $18,000 and save $5,000 per year after taxes. Energy costs are expected to escalate at an annual rate of 8% and the firm's discount rate is 10%. No salvage value is to be considered in either case.

Determine the savings to investment ratios for these energy-saving projects if both have a life of 10 years.

13
Computer Analysis

Computer programs are of great value in the analysis of problems which are complex (such as application of life cycle costing to building construction), and especially where the program can be used repeatedly (for example, weatherization retrofits).

For problems which are relatively uncomplicated and are not repetitive a computer program is not necessary. Preparing and utilizing computer programs can be costly, and they should be employed only if the savings justify their cost. The reader who has no use for computer programs may skip the following explanation and move forward to the section on sensitivity analysis.

To illustrate how computer programs can be used, a problem encountered earlier in this workbook is programmed in FORTRAN IV computer language. Although programming is not within the scope of this book, an effort has been made to make the programming process understandable to the uninitiated by means of this application. The purpose of this section is to demonstrate how a computer program can be utilized to perform a life cycle costing study and to facilitate sensitivity analysis.

METHODOLOGY

The steps to be taken in the development of the computer program are as follows:

Step One

State the total model ("model" is a synonym for "equation"). For example:

Computer Analysis

LCC = initial cost + present value of occasional costs + present value of terminal costs.

Step Two

State the models for determining each of the above costs. For example, one of the occasional costs is the annual maintenance cost. Using MC to represent this cost and ZMC to represent the present value of the maintenance costs, the model can be stated as follows:

$$ZMC = MC \times \frac{(1 + i)^n}{i(1 + i)^n}$$

where i = discount rate and n = years. The reader may recognize the factor on the right from an earlier section of this book on the time value of money.

Step Three

Accumualte the costs into the LCC figure using the model in Step One and print it. For example, if IC represents initial cost, and there is no salvage value, the basic model will be:

$$LCC = IC + ZMC$$

EXAMPLE

The air pollution problem (page 128) is solved here with a computer program. The steps outlined above are explained in this section and the actual program and printout of results appear in Appendix C.

Step One

Create a set of abbreviations for all the factors entering the problem: that is, costs, rates, etc. See the printout of the program in Appendix C for the abbreviations used for this problem. The abbreviations should be given in the program so that the reader can follow the reasoning.

The basic model appears near the end of the program:

$$TPWC = IC + ZEC + ZSC + ZMC + ZOC + ZOR$$

This statement reads as follows: The total present worth of costs (life cycle cost) equals the initial cost plus the present worth of energy costs, surveillance costs, maintenance costs, occasional repairs and occasional replacements.

Step Two

ZEC, ZSC, ZMC, ZOC, and ZOR must be calculated individually and the amounts will be entered into the basic model near the end of the program.

For example, ZEC, the present worth of the energy costs, equals energy cost times the discount/escalation present worth factor. The statement is: ZEC = EC × DEPW. The DEPW is calculated on the previous line which is a programming statement of the following (from Appendix A):

$$DEPW = \frac{\frac{1+ER}{1+i}\left[1 - \left(\frac{1+ER}{1+i}\right)^n\right]}{1 - \frac{1+ER}{1+i}}$$

Computation of ZOC is accomplished by obtaining the present worth of the escalated repair cost which occurs at the end of 7 years, and adding to this the present worth of the repair cost occurring at the end of every succeeding 7 years. Several equations are needed for this model, of which the key is:

$$ZOR = ZOR + OR \times \left(\frac{1+RR}{1+i}\right)^{YS}$$

Step Three

After each of the components of the life cycle cost has been calculated, the basic model appears in the program, i.e.

TPWC = IC + ZEC + ZSC + ZMC + ZOC + ZOR

The average annual cost is then found by multiplying the TPWC by the uniform capital recovery factor as follows:

$$AAC = TPWC \times \frac{i(1+i)^n}{(1+i)^n - 1}$$

Finally, all desired information obtained in the program is written out:

WRITE (6, 14) IC, ZEC, ZSC, ZMC, ZOC, TPWC, AAC

SENSITIVITY ANALYSIS

Sensitivity analysis is the determination of the effect on LCC of a change in one or more of the factors affecting total cost. These factors would include life expectancy, individual costs, salvage value, and escalation rates. The discount rate should not be manipulated for sensitivity analysis.

A computer program provides an important advantage for sensitivity analysis since the problem can be solved quickly, repeatedly, for the various combinations of the input factors. Usually, most of the input factors are estimates and it is helpful to know if small changes in those estimates will shift the LCC advantage between alternatives.

For example, solving the air pollution control problem using an energy escalation rate of 10% rather than 8% produces the following life cycle cost comparison:

Life Cycle Costs	*8%*	*10%*
Electrostatic Precipitator	29,219,920	31,674,992
Fabric Filter Bag Collector	26,140,496	27,840,320
Difference:	3,079,424	3,834,672

The conclusion to be reached, of course, is that even if the energy escalation rate exceeds the 8% estimate, the electrostatic precipitator is still more costly than the fabric filter bag collector.

LCPBM

The Life Cycle Planning and Budgeting Model (LCPBM) is a computer model used by the United States General Services Administration to perform life cycle costing analyses of the options available for satisfying space requirements. Costs are categorized as in UNIFORMAT (Appendix B) and entered into an economic model which evaluates the financial implications of the available options. Sensitivity analysis is performed as building, leasing, and renovation alternatives are compared using various model assumptions and cost estimates. The LCPBM does not automatically choose the most economical mode of acquisition, but it does provide the user with the information needed to make a professional judgment.

Readers interested in details about the LCPBM are referred to the "Final Report, Life Cycle Planning and Budgeting Model," Volume 1, Project Summary, July 1977, published by Public Buildings Service, General Services Administration, Washington, D.C.

SELF-STUDY PROBLEMS

Problem 13-1.—An airport authority is considering installation of a moving sidewalk and desires to perform an LCC analysis on it. Using the data given below, determine the present worth of all costs as well as an average annual cost for lifetimes of 10 to 30 years.

Equipment cost	$9,000,000
Structure cost	9,000,000
Operating and maintenance escalation rate	.07
Energy escalation rate	.07
Discount rate	.10
Current operating and maintenance costs	430,000
Annual KWH usage	9,500,000
Current energy cost	.03 KWH

Replacement escalation rate is .06.
Replacement costs: Year 15 $9,000,000
                   Year 30   2,000,000

The following will help in understanding the computer program provided in the Solutions section.

Step One:

   Model: TCOST = FCOST + RCOST + ECOST + OCOST
where
   FCOST = total first costs
   RCOST = present worth of replacement costs
   ECOST = present worth of energy costs
   OCOST = present worth of operating and maintenance costs
   TCOST = present worth of all costs

Step Two:

   FCOST = Equipment cost + structure cost

RCOST = Replacement costs (which are in today's dollars) escalated and then discounted. Using RR to represent escalation rate and DR the discount rate, we have RCOST $\times (1 + RR)^n/(1 + DR)^n$.

ECOST = Energy cost escalated and discounted. Current annual energy cost is cost per KWH $\times$ annual KWH usage, i.e. CK $\times$ KW. Since electric bills are paid throughout the year we will assume that an approximate equivalent is an annual payment after .6 of the year has elapsed. Therefore, the present worth of the energy cost for Year 1 is:

$$\text{ECYR1} = \text{CK} \times \text{KW} \times \frac{1 + \text{RE}(.6)}{1 + \text{DR}(.6)}$$

and

$$\text{ECOST} = \text{ECYR1} \times \frac{1 - \left(\frac{1 + \text{RE}}{1 + \text{DR}}\right)^n}{1 - \left(\frac{1 + \text{RE}}{1 + \text{DR}}\right)}$$

where RE = energy escalation rate.

OCOST = Operating and maintenance costs escalated and discounted. For Year 1:

$$\text{OCYR1} = \text{AO} \times \frac{1 + \text{RO}(.6)}{1 + \text{DR}(.6)}$$

where AO = current operating and maintenance cost

and RO = operating and maintenance escalation rate.

Then,

$$\text{OCOST} = \text{OCYR1} \times \frac{1 - \left(\frac{1 + \text{RO}}{1 + \text{DR}}\right)^n}{1 - \left(\frac{1 + \text{RO}}{1 + \text{DR}}\right)}$$

Step Three:

TCOST = FCOST + RCOST + ECOST + OCOST
Repeat for as many years as desired.
If average annual cost is needed:

$$\text{AVCST} = \text{TCOST} \times \frac{\text{DR} \times (1 + \text{DR})^n}{(1 + \text{DR})^n - 1}$$

14

Discount Rates

Much has been written on discount rates and the methods of determining them, but what is clear is that there is no universally accepted way of determining the discount rate(s) that should be used by any type of institution. Our objective in this section is not to discuss the theory of discount rates, but rather to explain some basic concepts about rates and to describe some possible ways of determining discount rates in the public and private sectors. Practitioners may wish to adopt their own variations or to make adjustments to the methods described.

RATES MATTER

It should be clear that the choice of a rate does matter and that the decision between competing projects may be heavily affected by the rate chosen for discounting. This point may be emphasized by an illustration in which various rates are used for compounding and discounting.

Item A has a first cost of $39,273 and annual costs of $2,500. Item B has a first cost of $60,000 and annual costs of $389. The projected life of each item is 20 years. Average annual costs are as follows:

Interest Rate	Average Annual Cost	
	A	B
2%	$4,902	$4,059
5%	5,671	5,203
6%	5,924	5,620
7%	6,207	6,052
8%	6,500	6,500

(continued)

Interest	Average Annual Cost	
Rate	A	B
10%	$7,113	$7,598
12%	7,758	8,422

At the 8% rate, A and B will have the same average annual cost. If a rate less than 8% is used, B will have the lower average annual cost while A will have the advantage at higher rates.

At the 8% rate both projects will have a present worth of $63,819. At lower rates B will have the lower present worth, and at higher rates A will have the lower present worth. The importance of the selection of rate should be obvious.

PRIVATE SECTOR

Although from the point of view of an economist the proper discount rate to be used in analysis of a project is the opportunity rate, or the rate of return available on projects of equivalent risk, such a rate is seldom possible to ascertain. Therefore it is suggested that the rate most commonly applied in the private sector, the firm's cost of capital, be used. The cost of capital is the rate at which the firm is financed. The weighted average cost of capital method is frequently used to obtain this rate.

Weighted Average Cost of Capital (WACC)

The weighted average cost of capital concept can be explained by a simple illustration. Let's say that a firm is financed 40% by debt and 60% by equity (owners' capital). The cost of the debt on an after-tax basis is 6% while equity costs 12%. To obtain the WACC, multiply the debt weight by the cost of the debt, the equity weight by the cost of the equity and add. Thus:

	Weight	X	Rate	=	Product
Debt	40%		6%		2.4%
Equity	60%		12%		7.2%
			WACC	=	9.6%

Determination of the weights and the appropriate rates is explained in the following example.

EXAMPLE

Miller Inc.'s balance sheet ($million):

Short-term Assets	30	Non-interest Liabilities	20
Long-term Assets	90	Interest-Bearing Liabilities	30
		Common Stock	40
		Retained Earnings	30
Total	120		120

Given: Miller's Federal income tax rate is 40%, interest on the debt is 10%, the market price of its common stock is $70 and its EPS (earnings per share) is $8.20. Determine the weighted average cost of capital.

SOLUTION

Step One:

Interest-bearing liabilities	30,000,000		Debt	30,000,000
Common stock	40,000,000	or		
			Equity	70,000,000
Retained earnings	30,000,000			
	100,000,000			100,000,000

Therefore, debt = 30% of the total and equity = 70%.

Step Two:

Debt rate: Since the objective is to obtain an after-tax cost of capital, the debt rate should be adjusted for the tax factor. If the tax rate is 40% of the pretax debt rate is multiplied by (100% − 40%) or 60%. The debt rate is 10% × 60% = .06.

Equity Rate: There are several ways of obtaining the equity rate. One is to add the dividend rate (dividend/stock price) to the estimated growth rate of earnings. A second method is to divide EPS by stock price. The equity rate is 8.20/70 = .117.

Step Three:

	Weight	×	Rate	=	Product
Debt	30%		.06		.018
Equity	70%		.117		.082
		WACC		=	.10

End of Problem

Marginal Cost of Capital (MCC)

The weighted average cost of capital as calculated above may be used as the discount rate in LCC analysis provided the new financing will not alter the proportions of debt and equity and that the cost rates will not change. If there are any changes in proportion or rates the marginal (or incremental) cost of capital is more appropriate.

The marginal cost of capital is calculated as follows:

Step One: Find the total annual cost of the financing after the new capital has been added.

Step Two: Subtract from this figure the total annual cost prior to the financing to obtain the amount of increase.

Step Three: Divide the increase by the amount of the new financing.

EXAMPLE

Miller Inc. (same firm as above) will finance a $10,000,000 pollution control project with a bond issue at a cost of 11%. Because of the heavier proportion of debt in the capitalization after the financing, the market price of the common stock is expected to drop to 66. Find the MCC.

Step One:

	Amount	×	Rate	=	Annual Cost
(a) New debt	10,000,000	×	.066		660,000
Old debt	30,000,000	×	.06		1,800,000
(b) Equity	70,000,000	×	.124		8,680,000
					11,140,000

(a) Cost of new debt is 60% of 11% = .066
(b) Cost of equity is 8.20/66 = .124

Step Two:

New annual cost	11,140,000
Old annual cost	10,000,000
	1,140,000

Step Three:

$$\frac{1{,}140{,}000}{10{,}000{,}000} = 11.4\% = \text{Marginal Cost of Capital}$$

End of Problem

In the above problem the $10,000,000 financing will cost 11.4% and this is the rate which should be used for the discount rate in analyzing the project. Operationally, however, the marginal cost of capital may be difficult to ascertain since the effect of the financing on the price of the stock is unknown. A company with little debt outstanding may find that increased leverage (i.e., debt to equity ratio) will both improve the stock price and reduce the weighted average cost of capital. A firm with a high debt/equity ratio may experience opposite effects as after-tax earnings become more vulnerable to cyclical developments. If management does not have reason to believe that the MCC will be higher or lower than the WACC, the latter should be used for LCC analysis.

The question often arises: Why not just use the debt rate if the investment will be financed by debt? The answer lies in the explanation of marginal cost of capital. The financing of any project has an impact on the company's overall cost of capital and even if the MCC cannot be calculated, the effect is still there. The only way of capturing the entire effect of financing a project is to use the cost of capital.

Risk Adjustment

Companies may use a discount rate higher than the cost of capital for analyzing revenue-producing projects. For example, a firm with a cost of capital of 11% may set a discount rate scale of 10% for low-risk projects, 12% for medium-risk projects, and 14% for high-risk projects. Projects are assigned to a risk category largely on the basis of subjective analysis.

Nonrevenue-producing projects may be similarly classified by risk. An HVAC system which will reduce fixed costs exclusively will have a very low risk and may warrant a discount rate lower than the weighted average cost of capital by reducing the firm's overall riskiness. On the other hand, if the system's benefits are variable, a higher discount rate is appropriate. If the management is reluctant to classify projects by risk category for application of a scale of discount rates, the authors suggest using the firm's cost of capital for life cycle costing.

Investment Rates

Before moving on to the section on public sector discount rates, a brief explanation of the components of rates is necessary.

A "real" rate is the basic rate for the time value of money. In the absence of inflation, a riskless investment would yield the real rate. Exactly what the real rate is, is not known but most estimates are in the neighborhood of 3%.

When inflation enters the picture, the return on a riskless investment increases accordingly. With no inflation and a real annual rate of 3%, an investment of $1 should amount to $1.03 in a year. If the inflation rate is 7%, the return should compensate for this rate as well as the time value of money; that is, it should be $1.03 × 1.07 = $1.1021 for a total yield of 10.21%. (Addition of the rates, i.e., 3% + 7% = 10%, is common, but multiplication will provide an accurate figure.)

The return on United States government securities is considered the riskless rate since there is little doubt that the principal and interest will be paid. It is composed of the real rate and the inflation rate, but since it fluctuates in response to monetary and fiscal policy, the rate on a particular day may not be appropriate for analytical purposes. The average over the course of the most recent business cycle would be a better estimate.

A gradation of risks prevails on investments throughout the economy and those risks are reflected in the spectrum of rates. A risky rate is composed of the real rate, the inflation rate, and a risk rate.

PUBLIC SECTOR

The discount rates used by the Federal Government are determined administratively. These rates are set by or in consultation with the Director of the U.S. Office of Management and Budget (OMB). The present rates are 7% for leasing or purchasing real property and 10% for other kinds of decisions (OMB Circulars A-94 and A-104). The 10% rate applies to all energy conservation projects in new, renovated or leased facilities.

The 10% rate dictated by OMB is a before-tax and after-inflation rate equivalent to the average in the private sector (see Stockfisch paper for the explanation and calculations on which the rate is based). Readers interested in the theory of why the government's discount rate should be the same as the private sector's are referred to the articles listed in the bibliography.

A philosophical discussion of what the Federal government's discount rate should be is not relevant to the objectives of this text but the authors feel that in fairness to the reader they should mention their strong dissatisfaction with OMB's method of determining the rate. Moreover, none of the methods for determining the rate which have been proposed are really satisfactory. Given the present state of confusion on the subject it is our opinion that the best choice of rate for the Federal government is an open market rate on its securities. This would not be a real rate, and allowance for inflation would have to be made, as described in this book.

State and Local Governments

As with the Federal government's discount rate, we find no acceptable theory of what the rates should be for state and local governments. Some of the current practices, however, (use of zero % or the borrowing rate) are very unsatisfactory, and we feel obliged to offer a method of obtaining a rate even though we will be a bit uncomfortable with it.

Use of a zero % discount rate was discussed in Chapter 2, and we hope that that notion has been laid to rest.

Use of the interest rate on bond issues is not acceptable either since interest on the issues of state and local governments is not subject to Federal income tax and therefore yield rates are lower than they would otherwise be. The tax-free status of the interest is an indirect type of social subsidy to the issuers and the yields are not truly reflective of the social cost.

It is recommended that for LCC analysis state and local governments use their long-term borrowing rate adjusted for the tax-exempt status of the interest payments. In other words, they should use the interest rate they would be paying if the interest were subject to Federal income tax.

The records show that the yields on high quality long-term tax-exempt bonds generally fluctuate in harmony with the yields of U.S. Government securities, though mostly 1 to 1½% below. Since U.S. Government bonds are considered riskless securities, the lower yields on tax-exempts can only be attributed to their tax-exempt status.

Approximation of what a state or local government yield should be is possible if we make two assumptions: (1) that the adjusted rates on the highest quality tax-exempts would be insignificantly higher than the rates on U.S. Government bonds and (2) that the prevailing differences in yields between higher and lower quality issues should apply equally in an adjusted spectrum of rates. Based on these assumptions, therefore, we recommend that for LCC discounting states and municipalities add 1½% to the yield rates on their long-term bonds, but with a minimum rate not less than the average yield rate on long-term U.S. bonds over the most recent business cycle.

SELF-STUDY PROBLEMS

Problem 14-1.—Find the after-tax weighted average cost of capital for the following case. The price of the common stock on the New York Stock Exchange is $60. The dividend is $4.80 and the estimated long-term annual growth rate of earnings is 6%. The interest rate on the long-term debt is 8%. The company has been paying 40% of its earnings to the IRS.

Debt	$40,000,000
Common Stock	20,000,000
Retained Earnings	20,000,000
	$80,000,000

Problem 14-2.—Find the marginal cost of capital for the company in Problem 14-1 if a planned expansion will be financed by $10,000,000 of new debt at 10% and $10,000,000 of equity at the existing rate.

Problem 14-3.—DEF Heating Corp. is bringing suit for $50,000 against an official of a large state because a contract awarded on the basis of lowest life-cycle bid went to a competitor, ABC Heating Co. DEF's equipment had an initial cost of $100,000 and estimated annual operating and maintenance costs of $9,000 over its 20-year life. ABC's equipment had an initial cost of $115,000 and annual costs of $8,000 over its 20-year life. The official had assumed that since no bonds would have to be issued to purchase the equipment the appropriate discount rate should be 0%. DEF maintains that a discount rate of at least 7% should have been used since that is the approximate current yield on bonds of that state. Determine the life cycle cost by both methods and comment on the likely outcome of the case.

Case Studies

CASE STUDY NO. 1

Imagine the enthusiasm with which a salesman of shower head fixtures was greeted when he made a proposal that all 100 shower heads in a just-completed dormitory of a state-owned educational institution be replaced with his model. He did manage to get a hearing, though, and his story is the basis for this case.

The salesman claimed that life cycle costing would prove that the state could save money by throwing away the existing shower heads and replacing them with his model. He provided the following data (some of which was estimated) on which he based his analysis.

Both heads have a 10-year life. Water costs $1.90 per 1,000 gallons and is heated by electricity to 100 degrees Fahrenheit. The average water supply is 55 degrees F. The present electric rate is $.04 per KWH.

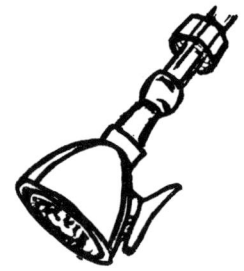

	New Head	*Existing Head*
Water consumption (gal./min.)	2.8	3.5
Average time per shower (min.)	6	6
No. showers/head/year	365	365
Cost of new installation	$26	0

Check the salesman's contention by answering the following questions:
  (1) What are the present worth values of the alternatives using a 10% discount rate?
  (2) Assuming an annual cost escalation rate of 8%, what is the payback period for the new shower head?

Case Studies

Reminder.—The annual cost of heating water calculations were discussed in Chapter 5.

$$\$ = \frac{\text{gals/yr} \times 8.34 \text{ lbs/gal} \times \Delta T}{\text{Eff.}} \times \frac{\$/\text{Fuel Unit}}{\text{Btu/Fuel Unit}}$$

or, when electricity is the source of heat:

$$\$ = \frac{\text{gals/yr} \times 8.34 \text{ lbs/gal} \times \Delta T}{1.00} \times \frac{\$/\text{KWH}}{3,416 \text{ Btu/KWH}}$$

SOLUTION (1)

New Head

    Initial Cost $26.00

    Annual Costs

    Water consumed
    = 2.8 gal./min. × 6 min./showers × 365 showers/yr
    = 6,132 gal./yr

Water cost = $\frac{\$1.90}{1,000 \text{ gal.}}$ × 6,132 gal/yr = $11.65

Electrical cost = $\frac{6,132 \times 8.34 \times 45}{3,416}$ × .04 = $26.95

                                                                           $38.60

                                                                                 $38.60

Existing Head

    Initial Cost $0.00

    Annual Costs

    Water consumed
    = 3.5 gal./min. × 6 min./shower × 365 days/yr
    = 7,665 gal./yr

Water cost = $\frac{\$1.90}{1,000 \text{ gal.}}$ × 7,665 gal./yr = $14.56

Electrical cost = $\frac{7,665 \times 8.34 \times 45}{3,416}$ × .04 = $33.68

                                                                           $48.24

                                                                             $48.24

PRESENT WORTH

New Head

    Initial Cost $26.00

Annual Costs $38.60 × (UPW; 10, 10)
$38.60 × 6.1445 $237.18
                       $263.18

Existing Head
   Initial Cost $ 0.00
   Annual Costs $48.24 × 6.1445 = $296.41
                                             $296.41

SOLUTION (2)

New Head
   Annual cost of electricity $26.95
   Annual cost of water   11.65
   Total annual cost $38.60

Old Head
   Annual cost of electricity $33.68
   Annual cost of water   14.56
   Total annual cost $48.24

Simple Payback

$$\frac{26 - 0}{48.24 - 38.60} = \frac{26.00}{9.64} = 2.70 \text{ years}$$

True Payback

$$k = \frac{1 + .08}{1 + .10} = .98$$

From the True Payback Chart (Figure 11-2):
True payback = 2.80 years

CASE STUDY NO. 2

A new heating system for an electronics manufacturing plant is being designed with consideration being given to its future energy usage. The design engineer has been asked by the owner of the plant to investigate the economics of adding an energy recovery wheel (heat wheel) to the system in an effort to reduce future energy costs.

The engineer calculates that if the wheel were installed, it would reclaim enough energy from the plant's exhaust air that the conventional heating equipment could be reduced in size. With a reduction in size, the initial cost of the conventional heating equipment will be reduced from $126,000 to $120,000. The installed cost of the wheel is $16,000.

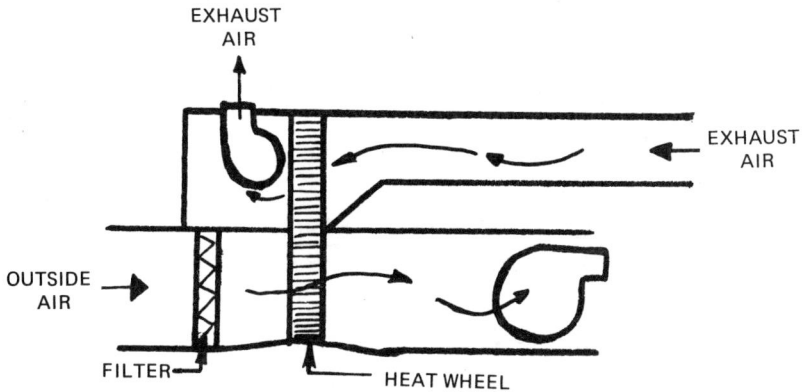

A listing of initial and annual costs for the two alternatives is as follows:

Initial Cost	Without Wheel	With Wheel
Conventional heating system	$126,000	$120,000
Wheel	—0—	16,000
	$126,000	$136,000
Annual Cost		
Energy cost	$ 30,000	$ 25,450
Maintenance of wheel	—0—	200
	$ 30,000	$ 25,650

The escalation rate is assumed to be 8% per year, the discount rate is 14%, and the life of the equipment is 20 years.

(1) Determine the simple payback for the installation of the energy recovery wheel.
(2) Determine the True Payback without considering taxes, i.e., discount-escalation payback.
(3) The design engineer determined that the wheel is eligible for the new 10% energy tax credit allowed by

172　Life Cycle Costing: A Practical Guide for Energy Managers

the 1978 Energy Tax Act. The firm's income tax rate is 46%. Determine the payback period for the wheel using this data. Straight-line depreciation is to be used.

(4) Assume the energy recovery wheel is to be installed as part of a combustion air system for a manufacturing process at the plant. Determine the payback.

SOLUTION (1)

Simple Payback

$$\frac{C}{S} = \frac{\$136,000 - \$126,000}{\$30,000/yr - \$25,650/yr}$$

$$\frac{C}{S} = 2.30$$

SOLUTION (2)

Discount-Escalation Payback

Determine the "k" factor:

$$k = \frac{1+e}{1+i} = \frac{1+.08}{1+.14} = .947$$

From the Discount-Escalation Payback nomograph
Payback = 2.53 years.

SOLUTION (3)

Initial Cost	Without Wheel	With Wheel
	$126,000	$136,000
10% Energy Tax Credit	—0—	(1,600)
	$126,000	$134,400
Difference		$ 8,400

Present Worth of Annual Costs

Year	Without Wheel	With Wheel	Savings With Wheel	Cumulative Savings
1	$12,805	$10,378	$2,427	$2,427
2	12,310	10,024	2,286	4,713
3	11,818	9,666	2,152	6,865
4	11,333	9,305	2,028	8,893

$$\text{Payback} = 3 + \frac{8{,}400 - 6{,}865}{8{,}893 - 6{,}865} = 3.76 \text{ years}$$

Supporting Calculations for Solution (3)

Energy Tax Credit $16,000 \times .10 = \$1,600$

Operating Cost
- With Wheel $25,650 \times .54 = \$13,851$
- Without Wheel $30,000 \times .54 = \$16,200$

Depreciation Tax Benefit
- With Wheel $136,000 \times 1/20 \times .46 = \$3,128$
- Without Wheel $126,000 \times 1/20 \times .46 = \$2,898$

Present Worth of Yearly Costs

With Wheel

Year 1 $\dfrac{13{,}851 \times 1.08 - 3{,}128}{1.14} = 10{,}378$

Year 2 $\dfrac{13{,}851 \times 1.08^2 - 3{,}128}{1.14^2} = 10{,}024$

Year 3 $\dfrac{13{,}851 \times 1.08^3 - 3{,}128}{1.14^3} = 9{,}666$

Year 4 $\dfrac{13{,}851 \times 1.08^4 - 3{,}128}{1.14^4} = 9{,}305$

Without Wheel

Year 1 $\dfrac{16{,}200 \times 1.08 - 2{,}898}{1.14} = 12{,}805$

Year 2 $\dfrac{16{,}200 \times 1.08^2 - 2{,}898}{1.14^2} = 12{,}310$

$$\text{Year 3} \quad \frac{16{,}200 \times 1.08^3 - 2{,}898}{1.14^3} = 11{,}818$$

$$\text{Year 4} \quad \frac{16{,}200 \times 1.08^4 - 2{,}898}{1.14^4} = 11{,}333$$

SOLUTION (4)

Initial cost will not reflect the additional 10% investment tax credit and the rest of the calculations will remain the same as in Solution (3):

Initial Cost	Without Wheel	With Wheel
	$126,000	$136,000
Energy Tax Credit	—0—	(1,600)
Investment Tax Credit	—0—	(1,600)
	$126,000	$132,800
Difference	$132,800 − $126,000 = $6,800	

Present Worth of Annual Costs are the same as in Solution (3).

$$\text{Payback} = 2 + \frac{6{,}800 - 4{,}713}{6{,}865 - 4{,}713} = 2.97$$

CASE STUDY NO. 3

The design architect for a high rise apartment building wants to use LCC to help him decide whether to install self-energizing exit signs or conventional electrically-powered exit signs in the apartment building. The self-energizing exit sign has a phosphorescent surface and requires NO electrical power to illuminate it. It has a 10-year life, after which time it must be reconditioned so that the phosphor will continue to emit the required illumination level. The cost of reconditioning the surfaces is $121.00 and this cost will escalate at 8% per year. Cost of labor to install the sign will be minimal since no electrical wiring is required.

The appropriate electrical rate to be used for the conventional exit sign is $.04/KWH. The life of the apartment building is 40 years, the assumed escalation rate is 8%, and the discount rate is 10%. The appropriate tax rate is 40%.

Case Studies

The architect has compiled the following data to help him in making his choice:

Self-Energized Exit Sign

    Initial Cost (including installation) $142.00
    Annual Costs   0
    Reconditioning Cost every 10 years $121.00

Conventional Exit Sign

    Initial Cost $101.65
    (cost includes exit sign, conduit, wiring, junction box and labor to install)
    Recurring Cost
        Electrical cost/yr —
        Two 15-watt light bulbs $10.50
        Lamp replacement cost every 4 years $2.10

Determine the present worth value of both the self-energized and the conventional exit sign. Given: (DEF 40, 10, 8) = 28.08099.

SOLUTION

Present Worth Values

  Self-Energized Exit Sign

    Initial Cost $142.00

    Replacement Cost

$$\$121.00 \times \frac{(1.08)^{10}}{(1.10)^{10}} \times .60 = 60.43$$

$$\$121.00 \times \frac{(1.08)^{20}}{(1.10)^{20}} \times .60 = 50.30$$

$$\$121.00 \times \frac{(1.08)^{30}}{(1.10)^{30}} \times .60 = \underline{41.87}$$

                                                                         $294.60

  Conventional Exit Sign

    Initial Cost $101.65
Electrical Cost $10.50 × 28.08099 × .60 = $176.91
*Lamp replacement $2.10 × 6.403826 × .60 =   8.07
                                                                          $286.63

CASE STUDY NO. 4

In an effort to save energy, a fast-food restaurant owner is considering replacing an existing kitchen exhaust hood with a double-wall exhaust hood at a cost of $6,000.

With the existing hood, all the exhaust make-up air is drawn from conditioned spaces from within the restaurant resulting in a waste of energy. With the double-wall exhaust hood, outside air would be brought into the kitchen grill area and would carry cooking odors and fumes outside through an exhaust duct. Concentric ducts would be installed with the outer duct supplying outside air and the inner duct used for exhausting that air. A simplified cross section of the hood is shown on the next page.

---

*$n = 9$,   Adj $i = \dfrac{(1 + .10)^4}{(1 + .08)^3} - 1 = .16$

Case Studies

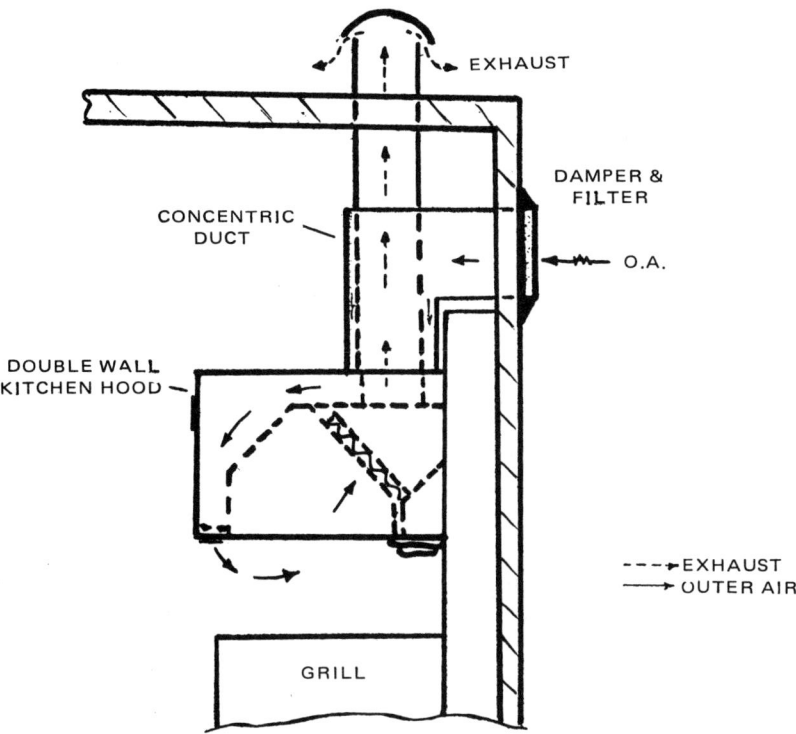

It is calculated that 5,000 CFM of conditioned air could be saved if this double-wall hood were installed. The restaurant is electrically heated and air-conditioned. A total of 6,200 heating and cooling degree days is applicable for the area. The exhaust hood is in continuous operation 14 hrs/day, 365 days/yr. Electricity costs $.04/KWH. A 10-year life is applicable for the exhaust hood.

From Chapter 5, we obtain the formula used to determine Heating Make-Up Air Costs (HMAC).

$$\text{HMAC} = \frac{1.08 \times \text{CFM} \times \text{DD} \times \text{hr/d}}{\text{Eff.}} \times \frac{\$/\text{Fuel Unit}}{\text{Btu/Fuel Unit}}$$

An escalation rate of 8%/yr for electricity costs is assumed and a discount rate of 10%/yr is to be used. The appropriate tax rate is 30%.

(1) Determine the True Payback period for the installation of the double-duct hood. Use straight-line depreciation.
(2) What is the Present Worth value of double-duct hood for its 10-year life?

SOLUTION

Savings in electrical costs per year (substitute in HMAC formula):

$$\text{HMAC} = \frac{1.08 \times 5000 \times 6200 \times 14}{1.0} \times \frac{\$.04/\text{KWH}}{3416 \text{ Btu/KWH}}$$

$$= \$5,489/\text{yr}$$

(1) Payback

Initial Cost	$ 6,000
Annual Costs	
Annual savings $5,489 × .70 =	(3,842)
Depreciation tax benefit:	
6,000 × 1/10 × .30 =	(180)
	$ 4,022

$$\text{Simple payback} = \frac{6,000}{4,022} = 1.49 \text{ yrs.}$$

True payback (using payback nomograph):

$$k = \frac{1 + .08}{1 + .10} = .98$$

True payback = 1.53 yrs.

(2) Present Worth

Initial Cost		$ 6,000
*Savings 4,022 × (DC 10, 10, 8)	=	
4,022 × 9.052914	=	(36,411)
		($30,411)

---

*Maintenance costs for both the existing hood and the double-wall hood are assumed to be equal. Therefore, these costs were ignored for the analysis.

CASE STUDY NO. 5

The decision must be made whether to install a heat-reclaim system in a student dining hall of a college located in northeastern United States.

The dining hall has two heating zones: a student dining area and a large kitchen. A 30,000 CFM electric resistance heating and ventilating unit supplies 100% outside air to both heating zones. This heated air is supplied 12 hours per day for 200 heating days each year. During the same period a total of 27,000 CFM of 80°F air is exhausted through four kitchen exhaust fans having individual capacities of 8,000, 8,000, 7,000 and 4,000 CFM.

What is the economic feasibility of installing a coil heat transfer loop system that would transfer heat normally exhausted by the four kitchen fans to the fresh air intake of the heating and ventilating unit via a pumped solution of glycol? The coil in the heating and ventilating unit would transfer the heat in the glycol solution to the incoming cool outside air and less electrical energy would be needed to raise the temperature of this air to the desired 70° F. A schematic diagram of the system is shown below:

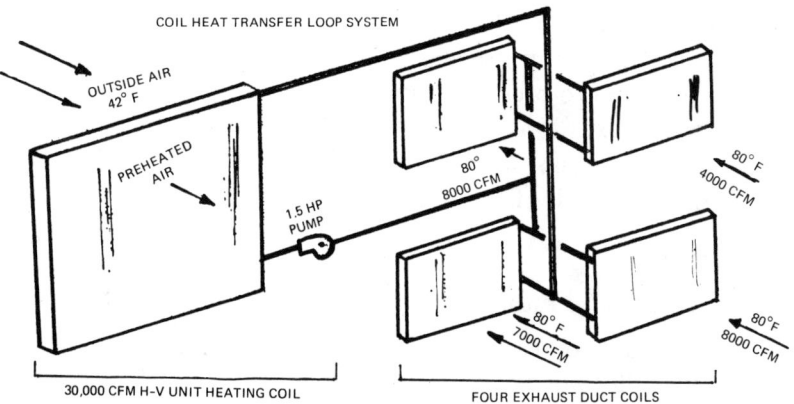

The following data are pertinent:
Initial cost: $22,000 for the system, $2,000 for the engineering design fee
Energy cost: Based on outside average temperature of 42 degrees during the heating season for that part of the

country, it was estimated that the annual savings in electrical energy would be 250,000 KWH. The local electricity cost was $0.34/KWH. The 1.5 HP pump needed for the system would consume an estimated 2,685 KWH/yr.

Maintenance: Labor at $8 an hour for 40 hours per year.
Taxes: Not relevant in this case.
Life: 20 years.
Terminal value: None
Discount rate: 10%
Escalation rates: Assumed 8% escalation rate for costs of electricity and labor.

SOLUTION 1

Neglecting escalation rates

Initial costs: 22,000 + 2,000	=	$ 24,000
Annual costs:		
Energy for pump 2,685 × .034/KWH	=	91
Labor $8 × 40 hours	=	320
Total annual costs	=	411
Annual electricity savings:		
250,000 KWH/yr × .034/KWH	=	(8,500)
Total annual savings	=	$ (8,089)
Converted to present worth:		
Initial cost:		$ 24,000
Annual savings: 8,089 × 8.51355		(68,866)
Present worth of total savings	=	$ (44,866)

SOLUTION 2

Allowing for escalation of costs

To allow for escalation of costs at 8%, multiply current costs by 1.08 to obtain the first year's costs, by 1.08^2 to obtain second year's costs, etc. Since the costs of electricity and labor are escalating at the same rate, the calculations are simplified by applying the discount/escalation factor to the net annual savings, i.e., $8,089.

Initial cost:			$ 24,000
Annual savings: 8,089 × 16.58821		=	(134,182)
Present worth of total savings		=	$ 110,182

PAYBACK

In Solution 1, where the annual savings are constant, the *simple payback* is calculated by dividing this figure into initial cost:

$$\$24{,}000 \div 8{,}089 = 2.97 \text{ years.}$$

To obtain the *discounted payback*:

Year			Year's Savings	Cumulative Savings
1	8,089 × .90909	=	7,354	7,354
2	8,089 × .82645	=	6,685	14,039
3	8,089 × .75132	=	6,077	20,116
4	8,089 × .68302	=	5,525	25,641

By interpolation, the discounted payback is determined to be

$$3 + \frac{24{,}000 - 20{,}116}{25{,}641 - 20{,}116} = 3.70 \text{ years.}$$

In other words, in 3.70 years the $24,000 initial cost and the cost of financing it at 10% interest will have been repaid.

True payback for Solution 2 (allowing for escalation) is obtained as follows:

Year			Year's Savings	Cumulative Savings
1	8,089 × .90909 × 1.08	=	7,942	7,942
2	8,089 × .82645 × 1.08^2	=	7,797	15,739
3	8,089 × .75132 × 1.08^3	=	7,655	23,394
4	8,089 × .68302 × 1.08^4	=	7,517	30,911

$$\text{True Payback} = 3 + \frac{24{,}000 - 23{,}394}{30{,}911 - 23{,}394} = 3.08 \text{ years.}$$

This method of calculating payback is the most meaningful of the three presented, and is recommended for all installations.

182   Life Cycle Costing: A Practical Guide for Energy Managers

CASE STUDY NO. 6

If the owner of the dining hall in Case Study No. 5 had been subject to taxation, calculate the life cycle costs.

The Business Energy Tax Act allows a 10% tax credit for investments in energy property placed in service after September 1978. Allowing for this credit, depreciation tax benefits (on a straight-line basis), tax deductibility of expenses based on 46% Federal income tax rate, and a property tax of $80 per $1,000 of assessed valuation which is 40% of cost, the life cycle cost is calculated as below. (The 10% investment tax credit would not apply in this case since the reclaim system was not related to a manufacturing process.) A 10% cost of money will be assumed for this tax-paying organization.

SOLUTION

Allowing for taxes.

Initial costs:
Engineering design fees	$ 2,000
System cost	22,000
10% energy credit	(2,400)
Total initial cost	$21,600

Annual costs and savings:
Energy and labor (91 + 320) × .54 =	222
Property tax 24,000 × 80/1000 × .40 × .54 =	415
Depreciation tax benefit 24,000 × 1/20 × .46 =	(552)
Electrical costs saved = 8,500 × .54	(4,590)

In calculating true payback, the labor and electricity costs are combined below since both are escalating at the same rate. The nonescalating costs and benefits also are grouped.

Escalating:

Electricity savings and labor = (4,590) − 222 = (4,368).

Nonescalating:

Depreciation tax benefit and property tax = (552) − 415 = (137).

Case Studies 183

Year			Year's Savings	Cumulative Savings
1	$\dfrac{(4{,}368 \times 1.08) + 137}{1.10}$	=	4,413	
2	$\dfrac{(4{,}368 \times 1.08^2) + 137}{1.10^2}$	=	4,324	8,737
3	$\dfrac{(4{,}368 \times 1.08^3) + 137}{1.10^3}$	=	4,237	12,974
4	$\dfrac{(4{,}368 \times 1.08^4) + 137}{1.10^4}$	=	4,153	17,124
5	$\dfrac{(4{,}368 \times 1.08^5) + 137}{1.10^5}$	=	4,070	21,197
6	$\dfrac{(4{,}368 \times 1.08^6) + 137}{1.10^6}$	=	3,990	25,187

$$\text{Payback} = 5 + \frac{21{,}600 - 21{,}197}{25{,}187 - 21{,}197} = 5.10$$

CASE STUDY NO. 7

A homeowner living in Central, PA wants to heat 60% of his hot water needs for his family of four by solar energy. The remaining 40% hot water requirements will be heated by a back-up electric hot water heater.

To help him decide whether the investment is worthwhile, he has compiled the following data:

Hot water required for a family of four, dishwasher and washing machine	68 gal/day
Hot water temperature	125° F
Cold water supply temperature	55° F
Present electric rate	$.06/KWH
Available Btu/sq ft/yr of collector for solar water heating, Harrisburg, PA.	198,000 Btu/sq ft/yr
Installed cost of solar water system	$60/sq ft

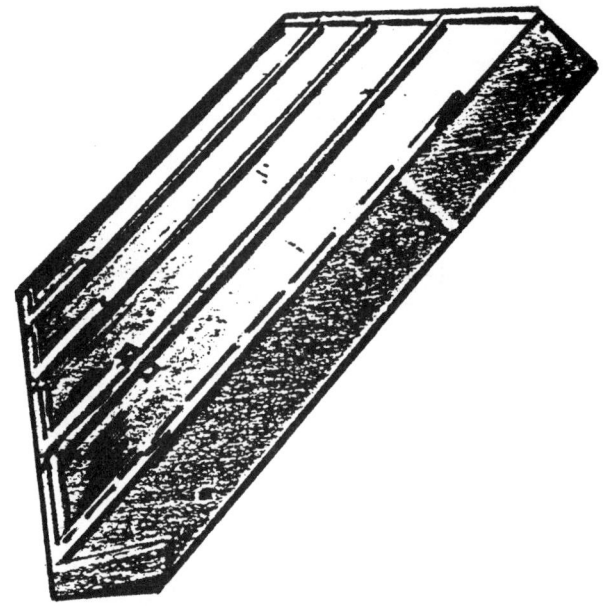

Taxes nonrefundable — 30% of first $2,000 and 20% of the next $8,000 (ref. 1978 National Energy Act).

He determines that he will owe at least $2,500 in taxes during the installation year.

Life of the solar equipment is 15 years. The appropriate interest rate for the project is 10% with electricity expected to escalate at 9% per year.

A HUD grant of $400 is available to the homeowner if he installs the solar equipment.

Should the project be implemented if the homeowner is to make his decision based on the net present worth saving that can be realized?

SOLUTION

Step 1 — Determine Btus Required
(Refer to Chapter 5: Water Heating Costs—WHC.)

Total Btu/yr = gal/yr × 8.34 × ΔT
 = 68 × 365 × 8.34 × (125−55)
Total Btu/yr = 14,489,916

Case Studies

Step 2 — Determine Solar Panel Size

Of the total Btu required, 60% will be contributed by solar panels.

$$\text{Required Btu from solar} = 14{,}489{,}916 \times .60$$
$$= 8{,}693{,}949 \text{ Btu/yr}$$

$$\text{Size of solar panels} = \frac{8{,}693{,}949 \text{ Btu/yr}}{198{,}000 \text{ Btu/sq ft/yr}}$$

$$\text{Size of solar panels} = 44 \text{ sq ft}$$

Step 3 — Determine Cost

Installed cost		
44 sq ft × $60/sq ft	=	$2,640
Less HUD grant	=	(400)*
Less tax credit		
(30% × 2,000)	=	(600)
(20% × 640)	=	(128)
Total Installed Cost	=	$1,512

Step 4 — Determine Savings

Saving in electrical energy:

$$\text{KWH} = \frac{\text{Btu/yr (solar)}}{3{,}416 \text{ Btu/KWH}}$$

$$= \frac{8{,}693{,}949}{3{,}416}$$

$$\text{KWH} = 2{,}545 \text{ saved}$$

Dollars saved per year:

$$\$ = \text{KWH} \times \frac{.06}{\text{KWH}}$$
$$= 2{,}545 \times .06$$
$$= \$153/\text{yr}$$

Step 5 — Determine Net Present Worth

Present worth of savings
for 15 years = Saving/yr × (DEF; 15, 10, 9)
= 153 × 13.954313 = $2,135

---
*HUD grant must be included in gross income for tax purposes.

Net Present Worth = P.W. savings—installation cost
= $2,135 − $1,512
Net Present Worth = $623

Conclusion: Based on the net present worth savings of $623, the homeowner should invest in the solar equipment.

Appendices

APPENDIX A

1. Derivation of Discount-Escalation Formula

Problem: To find present worth of a series of n annual payments growing at e% a year, discounted at i%.

Annual Payments:

Year 1	Year 2	Year 3	....	Year n
$1 × (1 + e)	$1 × (1 + e)²	$1 × (1 + e)³	....	$1 × (1 + e)ⁿ

Annual Payments Discounted:

Year 1	Year 2	Year 3	....	Year n
$1 × $\left[\dfrac{1+e}{1+i}\right]$	$1 × $\left[\dfrac{1+e}{1+i}\right]^2$	$1 × $\left[\dfrac{1+e}{1+i}\right]^3$	....	$1 × $\left[\dfrac{1+e}{1+i}\right]^n$

This is a geometric series. The formula for the sum of a geometric series is:

$$S = a\frac{(1 - r^n)}{1 - r}$$

where a = first term and r = multiplier.

In our problem, both a and r equal (1 + e)/(1 + i) and the sum of the series is the present worth of the series.

Therefore, P.W. = $\dfrac{\dfrac{1+e}{1+i}\left[1 - \left(\dfrac{1+e}{1+i}\right)^n\right]}{1 - \dfrac{1+e}{1+i}}$

2. Derivation of Adj. i

Assume a cost of operation, $A, occurs at equal intervals of p years. The number of payments over the life cycle is n and the annual escalation rate is e. With a discount rate of i, the present worth of the operating cost is:

Appendix A

$$\text{P.W.} = A\left(\frac{1+e}{1+i}\right)^P + A\left(\frac{1+e}{1+i}\right)^{2P} +$$

$$A\left(\frac{1+e}{1+i}\right)^{3P} + \ldots A\left(\frac{1+e}{1+i}\right)^{nP}$$

$$\text{P.W.} = A\left\{\left(\frac{1+e}{1+i}\right)^P + \left[\left(\frac{1+e}{1+i}\right)^P\right]^2 + \ldots \left[\left(\frac{1+e}{1+i}\right)^P\right]^n\right\}$$

Divide the numerator and denominator of each term $\left(\frac{1+e}{1+i}\right)^P$ by $(1+e)^{P-1}$.

$$\text{P.W.} = A\left[\frac{(1+e)}{\frac{(1+i)^P}{(1+e)^{P-1}}} + \frac{(1+e)^2}{\left(\frac{(1+i)^P}{(1+e)^{P-1}}\right)^2} + \ldots \frac{(1+e)^n}{\left(\frac{(1+i)^P}{(1+e)^{P-1}}\right)^n}\right]$$

This equation is similar to the formula for present worth of escalating costs in Chapter 10:

$$\text{P.W.} = A\left[\frac{1+e}{1+i} + \frac{(1+e)^2}{(1+i)^2} + \ldots \frac{(1+e)^n}{(1+i)^n}\right]$$

Equate the denominators using an adjusted i:

$$1 + \text{adj. } i = \frac{(1+i)^P}{(1+e)^{P-1}}$$

or

$$\boxed{\text{adj. } i = \frac{(1+i)^P}{(1+e)^{P-1}} - 1}$$

190   Life Cycle Costing: A Practical Guide for Energy Managers

3. Derivation of Formula for Logarithm Method of Calculating Discounted Payback

Let n = number of years for payback
  s = annual savings
  c = cost
  i = discount rate

We know that after n years the present value of the savings will equal the cost of the project, i.e., $c = s \times (UPW, n, i)$. We also know, from Chapter 2, that $UPW = \dfrac{(1+i)^n - 1}{i(1+i)^n}$.

Therefore: $c = s \times \dfrac{(1+i)^n - 1}{i(1+i)^n}$

and $i(1+i)^n = s \times \left[\dfrac{(1+i)^n - 1}{c}\right]$

$i(1+i)^n = \dfrac{s}{c}(1+i)^n - \dfrac{s}{c}$

$i(1+i)^n - \dfrac{s}{c}(1+i)^n = -\dfrac{s}{c}$

$(1+i)^n = \dfrac{-s/c}{i - s/c}$

$(1+i)^n = \dfrac{1}{1 - ci/s}$

Take the logarithm of both sides of the equation:

$$n \log(1+i) = \log \dfrac{1}{1 - ci/s}$$

Formula: $n = \dfrac{\log \dfrac{1}{1 - ci/s}}{\log(1+i)}$

4. Derivation of Formula for Logarithm Method of Calculating True Payback (Discount Escalation

Let n = number of years for payback
  c = cost
  S = annual savings

Appendix A

i = discount rate
e = annual escalation rate

We know that after "n" years, the present worth of the savings, including escalation, will equal the cost of the project.

or $\quad c = S\left(\dfrac{1+e}{1+i}\right) + S\left(\dfrac{1+e}{1+i}\right)^2 + S\left(\dfrac{1+e}{1+i}\right)^3 +$

$\qquad \ldots\ldots + S\left(\dfrac{1+e}{1+i}\right)^n$

Let $k = \dfrac{1+e}{1+i}$ Then:

(a) $\quad c = Sk + Sk^2 + Sk^3 + \ldots\ldots + Sk^n$

or

(b) $\quad c/k = S + Sk + Sk^2 + \ldots\ldots + Sk^{n-1}$

Subtracting equation (b) from equation (a)

$$c - \dfrac{c}{k} = Sk^n - S$$

Dividing by S

$$\dfrac{c}{S}\left(1 + \dfrac{1}{k}\right) = k^n - 1$$

$$k^n = \dfrac{c}{S}\left(1 + \dfrac{1}{k}\right) + 1$$

Taking log of equation

$$n \log k = \log\left[\dfrac{c}{S}\left(1 + \dfrac{1}{k}\right) + 1\right]$$

$$\boxed{n = \dfrac{\log\left[\dfrac{c}{S}\left(1 + \dfrac{1}{k}\right) + 1\right]}{\log k}}$$

APPENDIX B

UNIFORMAT

Categories	Related Periodic Costs
01 Foundations	Inspection, resealing, repair
02 Substructure	Repair, resealing, painting
03 Superstructure	Painting, refinishing, cleaning
04 Exterior Closure	Painting, cleaning windows, resealing replacements, seasonal installations
05 Roofing	Surface maintenance, cleaning of gutters, reflashing, repairs, repointing
06 Interior Construction	Cleaning, painting, replacements of carpeting and ceiling panels
07 Conveying Systems	Operating personnel, energy costs, costs of inspection, licensing, maintenance contract, cleaning, painting
08 Mechanical Systems (HVAC, plumbing, fire protection, etc.)	Operating personnel, energy costs, inspection, cleaning and maintenance, replacement of parts, painting
09 Electrical Systems	Operating personnel, energy costs, inspection, testing, maintenance, relamping, replacements
10 General Conditions	Costs of supervision, training, storage, reception, security, trash removal, extermination
11 Equipment	Operating personnel, inspection, maintenance contracts, cleaning, parts replacement, repairs, energy
12 Sitework	Groundskeepers, security, snow removal, cleaning, replacements, repairs, energy costs

Source Document:
 "Cost Control and Estimating System—Data Base"
 U.S. General Services Administration

APPENDIX C

```
      REAL IC, MC, MR, I
      INTEGER YR, YS
C     ABBREVIATIONS USED:
C        IC=INITIAL COST
C        EC=ANNUAL ENERGY COST
C        SC=SURVEILLANCE COST
C        MC=MAINTENANCE COST
C        OC=OCCASIONAL REPAIRS
C        OR=OCCASIONAL REPLACEMENTS
C        N=NUMBER OF YEARS LIFE
C        I=DISCOUNT RATE
C        ER=ENERGY ESCALATION RATE
C        MR=MAINTENANCE ESCALATION RATE
C        RR=REPLACEMENT ESCALATION RATE
C        YR=YEARS BETWEEN OCCASIONAL REPAIRS
C        YS=YEARS BETWEEN OCCASIONAL REPLACEMENTS
C        DEPW=DISCOUNT/ESCALATION P.W. FACTOR
C        Z PREFIX INDICATES P.W. OF A RECURRING COST
C        TPWC=TOTAL PRESENT WORTH OF COSTS
C        AAC=AVERAGE ANNUAL COST
    3   FORMAT (' ', 'INPUT INFORMATION',/,
    1    '   INITIAL COST', F24.0,/,
    2    '   ANNUAL ENERGY COST', F18.0,/,
    3    '   ANNUAL SURVEILLANCE COST', F12.0,/,
    4    '   ANNUAL MAINTENANCE COST', F13.0,/,
    5    '   OCCASIONAL REPAIRS', F16.0, '/, I1, 'YRS.',/,
    6    '   OCCASIONAL  REPLACEMENTS', F11.0,'/', I1,' YRS.',/,
    7    '   LIFE', 129,/,
    8    '   DISCOUNT RATE', F20.2,/,
    9    '   ENERGY ESCALATION RATE', F11.2,/,
    A    '   MAINTENANCE ESCALATION RATE', F6.2,/,
    B    '   REPLACEMENT ESCALATION RATE', F6.2,/)
    4   FORMAT ( ' ', 'SOLUTION',/,
    1 '   INITIAL COST', F28.0,/,
    2 '   ENERGY COST', F29.0,/,
    3 '   SURVEILLANCE COSTS', F22.0,/,
    4 '   MAINTENANCE COSTS', F23.0,/,
    5 '   OCC. REPAIR COSTS', F24.0,/,
    6 '   REPLACEMENT COSTS', F23.0,/,
    7 '   TOTAL PRESENT WORTH OF COSTS', F12.0,/,
```

```
      8 '   AVERAGE ANNUAL COST', F21.0)
      CHARACTER *50 TITLE
   10 FORMAT (A50)
      READ (5, 10) TITLE
      WRITE (6, *) TITLE
      READ (5, *) IC
      READ (5, *) EC
      READ (5, *) SC
      READ (5, *) MC
      READ (5, *) OC
      READ (5, *) OR
      READ (5, *) N
      READ (5, *) I
      READ (5, *) ER
      READ (5, *) MR
      READ (5, *) RR
      READ (5, *) YR
      READ (5, *) YS
      WRITE (6, 3) IC, EC, SC, MC, OC, YR, OR, YS, N, I, ER, MR,
                   RR
      DEPW = ( (1+ER) / (1+I) ) * (1-( (1+ER) / (1+I) ) ** N) / (1-
             ( (1+ER) / (1+I) ) )
      ZEC = EC * DEPW
      DEPW = ( (1 + MR) / (1+I) ) * (1-( (1+MR) / (1+I) ) ** N ) /
             (1-( (1+MR) / (1+I) ) )
      ZSC = SC * DEPW
      ZMC = MC * DEPW
      K = 1
      LR = YR
      ZOC = 0
   21 ZOC = ZOC + OC * ( (1+MR) / (1+I) ) ** YR
      K = K+1
      YR - LR*K
      IF (YR.GE.N) GO TO 20
      GO TO 21
   20 K = 1
      LZ = YS
      ZOR = 0
   23 ZOR = ZOR + OR * ( (1+RR) / (1+I) ) ** YS
      K = K + 1
      YS = LZ * K
```

Appendix C

```
    IF (YS. GE. N) GO TO 22
    GO TO 23
 22 TPWC = IC + ZEC + ZSC + ZMC + ZOC + ZOR
    AAC = TPWC * ( (I * ( (1+I) * * N) ) / ( (1+I) * * N−1) )
    WRITE (6, 4) IC, ZEC, ZSC, ZMC, ZOC, ZOR, TPWC, AAC
    STOP
    END
```

Analysis of Electrostatic Precipitator using energy escalation rates of 8% and 10%:

PROJECT: ELECTROSTATIC PRECIPITATOR	
INPUT INFORMATION	
INITIAL COST	14100000.
ANNUAL ENERGY COST	1145583.
ANNUAL SURVEILLANCE COST	5110.
ANNUAL MAINTENANCE COST	73800.
OCCASIONAL REPAIRS	480000. /7 YRS.
OCCASIONAL REPLACEMENTS	0. /2 YRS.
LIFE	20
DISCOUNT RATE	0.14
ENERGY ESCALATION RATE	0.08
MAINTENANCE ESCALATION RATE	0.08
REPLACEMENT ESCALATION RATE	0.08
SOLUTION	
INITIAL COST	14100000.
ENERGY COST	13627329.
SURVEILLANCE COSTS	60786.
MAINTENANCE COSTS	877891.
OCCASIONAL REPAIR COSTS	553926.
REPLACEMENT COSTS	0.
TOTAL PRESENT WORTH OF COSTS	29219920.
AVERAGE ANNUAL COST	4411801.
PROJECT: ELECTROSTATIC PRECIPITATOR	
INPUT INFORMATION	
INITIAL COST	14100000.
ANNUAL ENERGY COST	1145583.
ANNUAL SURVEILLANCE COST	5110.
ANNUAL MAINTENANCE COST	73800.
OCCASIONAL REPAIRS	480000. /7 YRS.
OCCASIONAL REPLACEMENTS	0. /2 YRS.

196   Life Cycle Costing: A Practical Guide for Energy Managers

LIFE	20
DISCOUNT RATE	0.14
ENERGY ESCALATION RATE	0.10
MAINTENANCE ESCALATION RATE	0.08
REPLACEMENT ESCALATION RATE	0.08

SOLUTION

INITIAL COST	14100000.
ENERGY COST	16082414.
SURVEILLANCE COSTS	60786.
MAINTENANCE COSTS	877891.
OCCASIONAL REPAIR COSTS	553926.
REPLACEMENT COSTS	0.
TOTAL PRESENT WORTH OF COSTS	31674992.
AVERAGE ANNUAL COST	4782482.

Analysis of Fabric Filter Bag Collector using energy escalation rates of 8% and 10%.

PROJECT: FABRIC FILTER BAG COLLECTOR
INPUT INFORMATION

INITIAL COST	12500000.
ANNUAL ENERGY COST	793168.
ANNUAL SURVEILLANCE COST	5110.
ANNUAL MAINTENANCE COST	50000.
OCCASIONAL REPAIRS	70000. /7 YRS.
OCCASIONAL REPLACEMENTS	636768. /2 YRS.
LIFE	20
DISCOUNT RATE	0.14
ENERGY ESCALATION RATE	0.08
MAINTENANCE ESCALATION RATE	0.08
REPLACEMENT ESCALATION RATE	0.08

SOLUTION

INITIAL COST	12500000.
ENERGY COST	9435162.
SURVEILLANCE COSTS	60786.
MAINTENANCE COSTS	594777.
OCCASIONAL REPAIR COSTS	80781.
REPLACEMENT COSTS	3469031.
TOTAL PRESENT WORTH OF COSTS	26140496.
AVERAGE ANNUAL COST	3946850.

Appendix C

```
PROJECT: FABRIC FILTER BAG COLLECTOR
INPUT INFORMATION
INITIAL COST                         12500000.
ANNUAL ENERGY COST                     793168.
ANNUAL SURVEILLANCE COST                 5110.
ANNUAL MAINTENANCE COST                 50000.
OCCASIONAL REPAIRS                      70000. /7 YRS.
OCCASIONAL REPLACEMENTS                636768. /2 YRS.
LIFE                                       20
DISCOUNT RATE                            0.14
ENERGY ESCALATION RATE                   0.10
MAINTENANCE ESCALATION RATE              0.08
REPLACEMENT ESCALATION RATE              0.08

SOLUTION
INITIAL COST                         12500000.
ENERGY COST                          11134990.
SURVEILLANCE COSTS                      60786.
MAINTENANCE COSTS                      594777.
OCCASIONAL REPAIR COSTS                 80781.
REPLACEMENT COSTS                     3469031.
TOTAL PRESENT WORTH OF COSTS         27840320.
AVERAGE ANNUAL COST                   4203500.
```

APPENDIX D

TAX BRIEFING

What follows is a brief synopsis of the authors' impressions of tax legislation relevant to investment as of the date of this workbook. For definitive explanations, see the appropriate publications of the Internal Revenue Service.

Homes

All tax credits listed under this section are nonrefundable. This means that the credits may be deducted from taxes due, but if the credits exceed the taxes the difference will not be refunded.

ENERGY CONSERVATION CREDITS

15% of cost up to maximum of $300. April 20, 1977 to December 31, 1985 on homes existing on April 20, 1977. Must be expected to remain in use for 3 years.

Allowed: Insulation, caulking, weather-stripping; exterior storm doors and windows; furnace burner replacements which reduce fuel usage; automatic flue dampers; automatic set-back thermometers or meters which display cost of energy usage.

Not allowed: Heat pumps; replacement boilers and furnaces; fluorescent lights; wood-burning stoves.

RENEWABLE ENERGY SOURCE CREDITS

30% of first $2,000 and 20% of next $8,000.

Allowed: Solar systems; wind or geothermal devices. On new or existing residences. Must be expected to remain in use for 5 years. Vacation homes do not qualify.

Schools, Hospitals

Federal grants of $900,000,000 over the next 3 years for energy audits and installation of energy conservation and solar energy measures. Will cover up to 50% of costs.

Appendix D

Business Credits (not utilities)

This section is divided into three parts: regular investment credits, energy investment credits, and special provisions.

REGULAR INVESTMENT CREDITS

Amount: 10%
Time limit: Permanent
Applies to: Depreciable or amortizable property having useful life of 3 or more years and included here:

(1) Tangible personal property.

(2) Tangible property used as integral part of (a) Manufacturing, (b) Extraction, (c) Production, (d) Furnishing of transportation, communications, electrical energy, gas, water or sewage disposal. Exception: Land and buildings and structural components.

(3) Elevators, escalators.

(4) Research facilities used in connection with (2) above.

(5) Storage facilities for certain commodities.

(6) Rehabilitation of commercial structures in use for at least 20 years and not rehabilitated in the last 20 years.

(7) Pollution control facilities amortized over 5 years except for 50% limit if financed with industrial development bonds.

(8) Commuter highway vehicles having useful life of 3 years.

Excluded:

(1) Property used to furnish lodging except hotels and motels.

(2) Certain air-conditioning or heating units fueled by oil or gas.

(3) Railroad rolling stock.

Early disposition:

Tax for year of disposal is increased by the difference between the credit originally allowed and what would have been allowed had the actual life been used.

Limitations on regular investment credit—

Amount of taxation to be offset:

1979 $25,000 + 60% of the tax liability over $25,000
1980 25,000 + 70% of the tax liability over $25,000
1981 25,000 + 80% of the tax liability over $25,000
After 1981 $25,000 + 90% of tax liability over $25,000.

Example: Investment is $1,000,000; credit is $100,000. Only in the current year can you get the credit. If your tax liability is $25,000, your credit is $25,000. If your tax liability is $80,000 your tax credit is $25,000 + 60% of (80,000 − 25,000) i.e., $58,000.

Basis: Cost of the new property including freight and installation. On used property the limit is $100,000.

> 100% if 7 years or over
> 2/3 if 5 years but less than 7
> 1/3 if 3 years but less than 5

ENERGY INVESTMENT CREDIT

Amount: 10%
Time limit: September 30, 1978 to June 2, 1983
Applies to: New equipment, including (a) boilers and burners which use an alternative to oil or gas; (b) geothermal equipment; (c) pollution equipment necessary for operation of above; (d) solar and wind energy equipment; (e) heat recovery, recycling, and synthetic fuel producing equipment.

Note: It may be possible for energy property to qualify for the energy investment credit even if it is considered a structural component.

Q: Is it possible for property to qualify for both the regular investment credit and the energy investment credit?
A: Yes. Property that meets both requirements is eligible for both credits.
Q: Is the energy investment credit refundable (see definition above)?
A: It is refundable for solar and wind energy property but not for other purposes. However, it can be carried forward 7 years or back 3 years to the extent that it exceeds the amount allowed in the current year.

Appendix D

SPECIAL PROVISIONS

Additional First-Year Depreciation

In addition to regular depreciation an additional depreciation of 20% of the cost of qualifying property, up to the limit, may be taken in the first year. The cost limit of the property is $10,000 for corporations. Salvage value is not used in the calculation.

Example: On July 1 the firm bought a used piledriver for $14,500. It had an estimated useful life of 10 years and a salvage value of $500.

Additional first-year depreciation—
20% of $10,000 = $2,000

Regular depreciation—
10% of $12,000
(subtract the $2,000 and the
$500 salvage) × ½ year =   600

Total depreciation = $2,600

Qualifying property—
New or used tangible personal property having a useful life of at least 6 years; must be purchased for use in business.

Nonqualifying property—
Land and buildings and structural components.

Salvage

When new or used personal property with useful life of 3 years is acquired, the salvage value may be reduced by an amount up to 10% of the full adjusted basis of the property when acquired.

Pollution Control Facilities

Three advantages for eligible PCFs: (a) Amortization over 60 months, (b) investment tax credit, (c) first-year depreciation. The amortization is based on the fraction of the property's basis attributable to its first 15 years. The investment tax credit is based on this amount also.

Glossary

Term	Definition
Alternatives	The different choices, propositions or methods by which objectives may be attained.
Annual Recurring Costs	Those costs which are incurred each year in an equal amount or in an amount that is increasing at a constant rate throughout the study period.
Annual Value	Past or future costs or benefits expressed as an equivalent uniform annual amount, taking into account the Time Value of Money.
Annuity	An annuity is a series of equal payments or receipts to be paid or received at the end of successive periods of equal time.
Base Year	The year to which all future and past costs are converted when a Present Value method is used.
Baseline	The baseline is the condition against which investment opportunities are measured.
Building Element	A single part, component or subsystem used in a building.
Constant Dollars	Constant dollars are dollar values expressed in terms of today's value. Constant dollars do not reflect cost escalation.
Compound Interest	Interest which is computed on both the original principal and its accrued interest.
Cost Effective	Estimated benefits (savings) from an energy conservation investment project are equal to or exceed the costs of the investment, where both are assessed over the life of the project.
Current Dollars	Current dollars are values which reflect cost escalation.

Glossary

Term	Definition
Depreciation	Depreciation is the allocation of the original cost of a facility or equipment to those time periods in which the asset is used.
Differential Cost	The difference in the total cost of two alternatives.
Discount Factor	A multiplicative number for converting costs and benefits occurring at different times to a common basis. Discount factors are obtained by solving a discount formula based upon one dollar of costs or benefits and the assumed Discount Rate.
Discount Rate	A rate used to relate present and future dollars. This is expressed as a percentage used to reduce the value of future ("tomorrow") dollars in relation to present ("today") dollars to account for the time value of money. It reflects the fact that dollars spent or received in the future are worth less than dollars spent or received in the present. The discount rate may be the interest rate or the desired rate of return.
Discounted Payback Period	The time required for the cumulative savings, net of future costs, from an investment to pay back the Investment Cost, considering the Time Value of Money.
Discounting	A technique for converting costs and benefits occurring over time to equivalent amounts at a common point in time.
Engineering Economic Analysis	A technique which allows the assessment of proposed engineering alternatives on the basis of considering

Term	Definition
	their economic consequences over time.
Economic Life	The economic life is defined as that period over which an investment is considered to be the lowest cost alternative for satisfying a particular need.
Equivalent Uniform Annual Cost	The total of all costs for a given decision or alternative, expressed as a uniform annual equivalent over the years in the analysis life cycle.
Financing Costs	Costs associated with financing capital investment in facilities. Includes both interest and one-time loan and finance charges.
Future Worth (or Value)	The future ("tomorrow") value of a present amount, given the time value of money.
Initial Capital Investment Costs	Costs associated with the initial planning, design and construction of a facility.
Interest Rate	The interest rate represents the annual time value of money and is referred to as the discount rate.
Life Cycle Costing	Life cycle costing is a method of expenditure evaluation which recognizes the sum total of all costs associated with the expenditure during the time it is in use. An evaluation technique.
Major Replacement Investment	Any significant future component replacement, included in the capital budget, which must be incurred during the study period in order to maintain the investment at a functional level.
Net Present Value of Investment Costs	The Present Value of the Initial Investment Cost plus the present

Term	Definition
	value of Major Replacement Investments less the present value of Salvage Values.
Net Present Value of Savings	The present value of life-cycle energy savings minus (or plus) the present value of the related increase (or decrease) in life-cycle costs.
Nonrecurring Cost	Cost not expected to be repeated at regular intervals.
Operation and Maintenance Costs	Costs incurred in the operation and maintenance of the facility itself. Sometimes termed "O&M" costs.
Opportunity Rate of Return	The effective annual rate of return an investor is able to obtain on an alternative investment opportunity. Thus, it may be used as the discount rate for the investment opportunity being analyzed.
Payback Period	The payback period is the length of time necessary to recover the initial investment of a project.
Present Value	Present value is the concept that a sum of money invested today will earn interest. It is based on the premise that a dollar today is worth more than a dollar to be received in the future by the amount of interest it earns.
Present Value Factor	The present value factor is the number by which an investment cost or a benefit value realized in the future is multiplied to yield its value in today's terms.
Present Worth	Synonymous with Present Value.
Project Development Period	The time period in which the facility is initially planned, designed and constructed.
Rate of Return	The rate of return is the interest rate which, over a period of time,

Term	Definition
	equates the benefits derived from an opportunity to the investment cost of the project.
Recurring Costs	Recurring costs are those costs which recur on a periodic basis throughout the life of a project.
Repair and Replacement Costs	Costs associated with restoring a facility to approximate its original performance.
Salvage Costs (or Values)	The residual costs or values of assets after they have provided the intended service within, or at the end of, the analysis life cycle. Usually expressed as a negative cost in the analysis.
Savings-To-Investment Ratio (SIR)	Either the ratio of Present Value savings to present value investment costs, or the ratio of Annual Value savings to annual value investment costs.
Sensitivity Analysis	Testing the outcome of an evaluation by altering one or more system parameters from the initially assumed values.
Single Compound Amount (SCA)	A discount factor used to convert a present sum of money to its future worth, given a discount (interest) rate and a length of time.
Single Present Worth (SPW)	A discount factor used to convert a future sum of money to its pres-worth, given a discount (interest) rate and a length of time.
Sunk Cost	A sunk cost is a cost which has already been made and should not be considered in measuring the economic performance of an investment alternative.
Time Horizon	The ending point of the life cycle cost analysis. The cutoff, or last year, of the analysis.

Glossary

Term	Definition
Time Value of Money	The time value of money is the difference between the value of a dollar today and its value at some future point in time if invested at a stated rate of interest.
Total Present Worth Cost	The total of all costs for a given decision or alternative, expressed in "today" dollars. All costs are brought to their present worth in the baseline year.
Uniform Capital Recovery (UCR)	A discount factor used to convert a present sum of money to a series of uniform annual sums of money, given a discount (interest) rate and a length of time.
Uniform Compound Amount (UCA)	A discount factor used to convert a series of uniform annual sums of money to their future worth, given a discount (interest) rate and a length of time.
Uniform Present Worth (UPW)	A discount factor used to convert a series of uniform annual sums of money to their present worth, given a discount (interest) rate and a length of time.
Uniform Sinking Fund (USF)	A discount factor used to convert a future sum of money into a series of uniform annual sums of money, given a discount (interest) rate and a length of time.
Useful Life	The period of time over which a building element may be expected to give service without major renewal. It may represent physical lifespan or time before technological obsolescence.
Value Analysis (or Engineering, or Management)	Any technique for evaluating a product in order to find a substitute at a lesser cost. Once a sub-

Term	*Definition*
	stitute solution is identified, its analysis in relation to the existing approach may suggest the use of life cycle cost analysis technique.
Zero Year	The zero year represents the day before the system being evaluated starts yielding economic benefits.

Bibliography

For information pertaining to the role of life cycle costing in the design and development of a system or product see *Design and Manage to Life Cycle Cost,* Benjamin S. Blanchard, M/A Press, Portland, Oregon, 1978.

The Life Cycle Planning and Budgeting Model (LCPBM) developed for the U.S. Public Buildings Service is a computer model which provides comparative LCC estimates for the options available for satisfying a particular space requirement. See *The Life Cycle Planning and Budgeting Model,* Vols. I-V, General Services Administration, Washington, D.C., 1977.

Information about life cycle costing application in health facilities has been developed under the sponsorship of the U.S. Public Health Service. See *Life Cycle Budgeting and Costing As An Aid in Decision Making,* Vols. I-V, Department of Health, Education and Welfare, Washington, D.C., 1977.

Architects and builders may find helpful the manual developed by the A.I.A.: *Life Cycle Cost Analysis: A Guide for Architects,* The American Institute of Architects, Washington, D.C., 1977.

Recommendations for energy savings in existing buildings can be found in two publications of the Federal Energy Administration:

(1) "Guidelines for Saving Energy in Existing Buildings," Building Owners and Operators Manual, ECM 1, 1975.

(2) "Guidelines for Saving Energy in Existing Buildings," Engineers, Architects and Operators Manual, ECM 2, 1975.

A worthwhile source on building energy use and load estimating is the *Handbook of Air Conditioning Systems Design,* Carrier Air Conditioning Co., McGraw-Hill, 1965.

Another good source on building energy use and load estimating is the *Handbook of Fundamentals, ASHRAE,* American Society of Heating, Refrigerating and Air Conditioning Engineers, Inc., 1977.

The U.S. Department of Commerce, National Bureau of Standards, has two publications with numerous recommendations for saving energy in industrial and commercial buildings: *Energy Conservation Program Guide for Industry and Commerce,* NBS Handbook 115 and NBS Handbook 115, Supplement 1, 1974.

For industrial and commercial energy consumption calculations see *Fan Engineering*, Buffalo Forge Co., 1970.

Five worthwhile articles on governmental discount rates:

(1) Arrow, K. J., "Discounting and Public Investment Criteria," in A. V. Kneese and S. C. Smith, Eds., *Water Research*, Baltimore, 1966.

(2) Bailey, M. J. and Jensen, M. C., "Risk and the Discount Rate for Public Investment," in M. C. Jensen, Ed., *Studies in the Theory of Capital Markets*, New York, Praeger, 1972.

(3) Baumol, W. J., "The Social Rate of Discount," *American Economic Review*, LVIII (Sept. 1968), pp. 788–802.

(4) Gordon, M. J., "A Portfolio Theory of the Social Discount Rate and the Public Debt," *Journal of Finance*, Vol. XXXI (May 1976) pp. 199–214.

(5) Stockfisch, J. A., "Measuring the Opportunity Cost of Government Investment," Research Paper P-490, Institute for Defense Analyses, March 1969.

Ranking methods for energy conservation projects are discussed at length in *Life Cycle Costing, A Guide for Selecting Energy Conservation Projects for Public Buildings*, NBS Building Science Series 113, U.S. Department of Commerce, National Bureau of Standards, 1978.

Cost estimation is covered in depth in *Cost Estimating for Engineering and Management*, P. F. Ostwald, Prentice Hall, 1974.

The Department of Energy's proposed methodology for life cycle cost analysis of Federal buildings, including energy price projections to 1995 was published in the *Federal Register*, April 30, 1979, pp. 25366–25382.

TIME VALUE

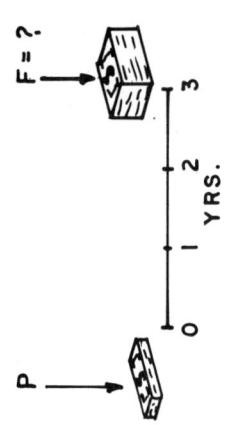

SINGLE COMPOUND AMOUNT (SCA) TABLE

IF YOU KNOW P (PRESENT WORTH) AND WANT TO FIND F (FUTURE WORTH), THEN:

$$P \times SCA = F$$

SINGLE PRESENT WORTH (SPW) TABLE

IF YOU KNOW F (FUTURE WORTH) AND WANT TO FIND P (PRESENT WORTH), THEN:

$$P = F \times SPW$$

UNIFORM CAPITAL RECOVERY (UCR) TABLE

IF YOU KNOW P (PRESENT WORTH OF MONEY) AND WANT TO FIND A (UNIFORM ANNUAL PAYMENT), THEN:

$$P \times UCR = A$$

OF MONEY

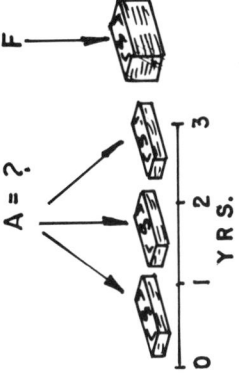

UNIFORM PRESENT WORTH (UPW) TABLE

IF YOU KNOW A (UNIFORM ANNUAL PAYMENT) AND WANT TO FIND P (PRESENT WORTH OF ALL THESE PAYMENTS), THEN:

$$P = UPW \times A$$

UNIFORM SINKING FUND (USF) TABLE

IF YOU KNOW F (THE FUTURE WORTH OF A SERIES OF ANNUAL PAYMENTS) AND WANT TO FIND A (VALUE OF THOSE ANNUAL PAYMENTS), THEN:

$$A = F \times USF$$

UNIFORM COMPOUND AMOUNT (UCA) TABLE

IF YOU KNOW A (UNIFORM ANNUAL PAYMENT) AND WANT TO FIND F (THE FUTURE WORTH OF THESE PAYMENTS, THEN:

$$A \times UCA = F$$

Interest Tables

2.00% Compound Interest Factors

Periods	Single Payment		Uniform Series			
	Compound Amount Factor SCA	Present Worth Factor SPW	Capitol Recovery Factor UCR	Present Worth Factor UPW	Sinking Fund Factor USF	Compound Amount Factor UCA
	1	2	3	4	5	6
1	1.02000	0.98039	1.02002	0.98037	1.00002	0.9999
2	1.04040	0.96117	0.51507	1.94150	0.49507	2.0199
3	1.06121	0.94232	0.34677	2.88379	0.32677	3.0602
4	1.08243	0.92385	0.26263	3.80763	0.24263	4.1214
5	1.10408	0.90573	0.21216	4.71332	0.19216	5.2038
6	1.12616	0.88797	0.17853	5.60129	0.15853	6.3079
7	1.14868	0.87056	0.15452	6.47184	0.13452	7.4340
8	1.17165	0.85349	0.13651	7.32531	0.11651	8.5827
9	1.19509	0.83676	0.12252	8.16203	0.10252	9.7543
10	1.21899	0.82035	0.11133	8.98238	0.09133	10.9494
11	1.24337	0.80427	0.10218	9.78662	0.08218	12.1683
12	1.26823	0.78850	0.09456	10.57511	0.07456	13.4117
13	1.29360	0.77304	0.08812	11.34812	0.06812	14.6799
14	1.31947	0.75788	0.08260	12.10599	0.06260	15.9734
15	1.34586	0.74302	0.07783	12.84898	0.05783	17.2928
16	1.37277	0.72845	0.07365	13.57741	0.05365	18.6387
17	1.40023	0.71417	0.06997	14.29157	0.04997	20.0114
18	1.42823	0.70017	0.06670	14.99172	0.04670	21.4116
19	1.45680	0.68644	0.06378	15.67815	0.04378	22.8398
20	1.48593	0.67298	0.06116	16.35109	0.04116	24.2966
21	1.51565	0.65978	0.05879	17.01086	0.03879	25.7825
22	1.54596	0.64685	0.05663	17.65768	0.03663	27.2981
23	1.57688	0.63416	0.05467	18.29184	0.03467	28.8440
24	1.60842	0.62173	0.05287	18.91356	0.03287	30.4209
25	1.64059	0.60954	0.05122	19.52307	0.03122	32.0293
26	1.67340	0.59759	0.04970	20.12064	0.02970	33.6698
27	1.70686	0.58587	0.04829	20.70650	0.02829	35.3431
28	1.74100	0.57438	0.04699	21.28087	0.02699	37.0500
29	1.77582	0.56312	0.04578	21.84396	0.02578	38.7909
30	1.81133	0.55208	0.04465	22.39604	0.02465	40.5667
31	1.84765	0.54125	0.04360	22.93729	0.02360	42.3780
32	1.88451	0.53064	0.04261	23.46791	0.02261	44.2255
33	1.92220	0.52024	0.04169	23.98813	0.02169	46.1100
34	1.96064	0.51004	0.04082	24.49815	0.02082	48.0321
35	1.99986	0.50004	0.04000	24.99818	0.02000	49.9927
36	2.03985	0.49023	0.03923	25.48840	0.01923	51.9925
37	2.08065	0.48062	0.03851	25.96901	0.01851	54.0323
38	2.12226	0.47120	0.03782	26.44020	0.01782	56.1129
39	2.16470	0.46196	0.03717	26.90215	0.01717	58.2351
40	2.20800	0.45290	0.03656	27.35503	0.01656	60.3998

3.00% Compound Interest Factors

Periods	Single Payment		Uniform Series			
	Compound Amount Factor SCA 1	Present Worth Factor SPW 2	Capitol Recovery Factor UCR 3	Present Worth Factor UPW 4	Sinking Fund Factor USF 5	Compound Amount Factor UCA 6
1	1.03000	0.97087	1.03001	0.97087	1.00001	0.9999
2	1.06090	0.94260	0.52262	1.91343	0.49262	2.0299
3	1.09273	0.91514	0.35353	2.82858	0.32353	3.0908
4	1.12551	0.88849	0.26903	3.71706	0.23903	4.1835
5	1.15927	0.86261	0.21836	4.57966	0.18836	5.3090
6	1.19405	0.83749	0.18460	5.41714	0.15460	6.4683
7	1.22987	0.81309	0.16051	6.23022	0.13051	7.6623
8	1.26677	0.78941	0.14246	7.01962	0.11246	8.8922
9	1.30477	0.76642	0.12844	7.78604	0.09844	10.1589
10	1.34391	0.74410	0.11723	8.53013	0.08723	11.4637
11	1.38423	0.72242	0.10808	9.25255	0.07808	12.8076
12	1.42576	0.70138	0.10046	9.95392	0.07046	14.1918
13	1.46853	0.68095	0.09403	10.63488	0.06403	15.6176
14	1.51258	0.66112	0.08853	11.29599	0.05853	17.0861
15	1.55796	0.64186	0.08377	11.93784	0.05377	18.5986
16	1.60470	0.62317	0.07961	12.56101	0.04961	20.1566
17	1.65284	0.60502	0.07595	13.16603	0.04595	21.7613
18	1.70243	0.58740	0.07271	13.75343	0.04271	23.4141
19	1.75350	0.57029	0.06981	14.32371	0.03981	25.1165
20	1.80610	0.55368	0.06722	14.87739	0.03722	26.8700
21	1.86028	0.53755	0.06487	15.41492	0.03487	28.6761
22	1.91609	0.52190	0.06275	15.93682	0.03275	30.5363
23	1.97357	0.50669	0.06081	16.44350	0.03081	32.4524
24	2.03278	0.49194	0.05905	16.93542	0.02905	34.4260
25	2.09376	0.47761	0.05743	17.41304	0.02743	36.4588
26	2.15658	0.46370	0.05594	17.87674	0.02594	38.5525
27	2.22127	0.45019	0.05456	18.36292	0.02456	40.7091
28	2.28791	0.43708	0.05329	18.76399	0.02329	42.9303
29	2.35655	0.42435	0.05211	19.18834	0.02211	45.2182
30	2.42724	0.41199	0.05102	19.60033	0.02102	47.5747
31	2.50006	0.39999	0.05000	20.00032	0.02000	50.0020
32	2.57506	0.38834	0.04905	20.38866	0.01905	52.5020
33	2.65231	0.37703	0.04816	20.76569	0.01816	55.0770
34	2.73188	0.36605	0.04732	21.13173	0.01732	57.7293
35	2.81384	0.35539	0.04654	21.48711	0.01654	60.4612
36	2.89825	0.34504	0.04580	21.83214	0.01580	63.2750
37	2.98520	0.33499	0.04511	22.16713	0.01511	66.1732
38	3.07475	0.32523	0.04446	22.49236	0.01446	69,1584
39	3.16699	0.31576	0.04384	22.80811	0.01384	72.2331
40	3.26200	0.30656	0.04326	23.11467	0.01326	75.4001

4.00% Compound Interest Factors

Periods	Single Payment		Uniform Series			
	Compound Amount Factor SCA 1	Present Worth Factor SPW 2	Capitol Recovery Factor UCR 3	Present Worth Factor UPW 4	Sinking Fund Factor USF 5	Compound Amount Factor UCA 6
1	1.04000	0.96154	1.04000	0.96154	1.00000	1.0000
2	1.08160	0.92456	0.53020	1.88608	0.49020	2.0399
3	1.12486	0.88900	0.36035	2.77508	0.32035	3.1215
4	1.16986	0.85480	0.27549	3.62988	0.23549	4.2464
5	1.21665	0.82193	0.22463	4.45181	0.18463	5.4163
6	1.26532	0.79032	0.19076	5.24212	0.15076	6.6329
7	1.31593	0.75992	0.16661	6.00205	0.12661	7.8982
8	1.36857	0.73069	0.14853	6.73274	0.10853	9.2142
9	1.42331	0.70259	0.13449	7.43532	0.09449	10.5827
10	1.48024	0.67556	0.12329	8.11089	0.08329	12.0061
11	1.53945	0.64958	0.11415	8.76046	0.07415	13.4863
12	1.60103	0.62460	0.10655	9.38508	0.06655	15.0257
13	1.66507	0.60057	0.10014	9.98564	0.06014	16.6268
14	1.73168	0.57748	0.09467	10.56312	0.05467	18.2918
15	1.80094	0.55526	0.08994	11.11839	0.04994	20.0235
16	1.87298	0.53391	0.08582	11.65230	0.04582	21.8244
17	1.94790	0.51337	0.08220	12.16567	0.04220	23.6974
18	2.02582	0.49363	0.07899	12.65930	0.03899	25.6453
19	2.10685	0.47464	0.07614	13.13394	0.03614	27.6711
20	2.19112	0.45639	0.07358	13.59033	0.03358	29.7780
21	2.27877	0.43883	0.07128	14.02916	0.03128	31.9691
22	2.36992	0.42196	0.06920	14.45112	0.02920	34.2478
23	2.46471	0.40573	0.06731	14.85685	0.02731	36.6178
24	2.56330	0.39012	0.06559	15.24695	0.02559	39.0825
25	2.66583	0.37512	0.06401	15.62209	0.02401	41.6458
26	2.77247	0.36069	0.06257	15.98278	0.02257	44.3116
27	2.88337	0.34682	0.06124	16.32957	0.02124	47.0841
28	2.99870	0.33348	0.06001	16.66306	0.02001	49.9675
29	3.11865	0.32065	0.05888	16.98370	0.01888	52.9661
30	3.24339	0.30832	0.05783	17.29202	0.01783	56.0848
31	3.37313	0.29646	0.05686	17.58849	0.01686	59.3282
32	3.50805	0.28506	0.05595	17.87354	0.01595	62.7013
33	3.64838	0.27409	0.05510	18.14764	0.01510	66.2094
34	3.79431	0.26355	0.05431	18.41118	0.01431	69.8578
35	3.94608	0.25342	0.05358	18.66460	0.01358	73.6520
36	4.10393	0.24367	0.05289	18.90826	0.01289	77.5981
37	4.26808	0.23430	0.05224	19.14256	0.01224	81.7021
38	4.43881	0.22529	0.05163	19.36786	0.01163	85.9701
39	4.61636	0.21662	0.05106	19.58447	0.01106	90.4089
40	4.80101	0.20829	0.05052	19.79276	0.01052	95.0253

5.00% Compound Interest Factors

Periods	Single Payment		Uniform Series			
	Compound Amount Factor SCA	Present Worth Factor SPW	Capitol Recovery Factor UCR	Present Worth Factor UPW	Sinking Fund Factor USF	Compound Amount Factor UCA
	1	2	3	4	5	6
1	1.05000	0.95238	1.05001	0.95237	1.00002	0.9999
2	1.10250	0.90703	0.53781	1.85938	0.48781	2.0499
3	1.15762	0.86384	0.36721	2.72321	0.31721	3.1524
4	1.21550	0.82271	0.28202	3.54589	0.23202	4.3100
5	1.27628	0.78353	0.23098	4.32942	0.18098	5.5255
6	1.34009	0.74622	0.19702	5.07563	0.14702	6.8018
7	1.40709	0.71069	0.17282	5.78630	0.12282	8.1418
8	1.47745	0.67684	0.15472	6.46313	0.10472	9.5489
9	1.55132	0.64461	0.14069	7.10773	0.09069	11.0263
10	1.62888	0.61392	0.12951	7.72165	0.07951	12.5776
11	1.71033	0.58468	0.12039	8.30632	0.07039	14.2065
12	1.79584	0.55684	0.11283	8.86315	0.06283	15.9168
13	1.88563	0.53033	0.10646	9.39347	0.05646	17.7126
14	1.97991	0.50507	0.10102	9.89854	0.05103	19.5982
15	2.07891	0.48102	0.09634	10.37956	0.04634	21.5780
16	2.18285	0.45812	0.09227	10.83767	0.04227	23.6569
17	2.29199	0.43630	0.08870	11.27396	0.03870	25.8397
18	2.40659	0.41553	0.08555	11.68948	0.03555	28.1317
19	2.52691	0.39574	0.08275	12.08522	0.03275	30.5382
20	2.65326	0.37690	0.08024	12.46210	0.03024	33.0651
21	2.78592	0.35895	0.07800	12.82105	0.02800	35.7183
22	2.92521	0.34186	0.07597	13.16290	0.02597	38.5042
23	3.07147	0.32558	0.07414	13.48847	0.02414	41.4294
24	3.22504	0.31007	0.07247	13.79854	0.02247	44.5008
25	3.38629	0.29531	0.07095	14.09385	0.02095	47.7258
26	3.55560	0.28125	0.06956	14.37508	0.01956	51.1120
27	3.73338	0.26785	0.06829	14.64293	0.01829	54.6676
28	3.92005	0.25510	0.06712	14.89802	0.01712	58.4009
29	4.11605	0.24295	0.06605	15.14098	0.01605	62.3209
30	4.32185	0.23138	0.06505	15.37237	0.01505	66.4369
31	4.53794	0.22036	0.06413	15.59272	0.01413	70.7587
32	4.76483	0.20987	0.06328	15.80259	0.01328	75.2966
33	5.00307	0.19988	0.06249	16.00244	0.01249	80.0613
34	5.25322	0.19036	0.06176	16.19281	0.01176	85.0643
35	5.51587	0.18129	0.06107	16.37410	0.01107	90.3174
36	5.79166	0.17266	0.06043	16.54675	0.01043	95.8332
37	6.08124	0.16444	0.05984	16.71120	0.00984	101.6248
38	6.38530	0.15661	0.05928	16.86780	0.00928	107.7060
39	6.70456	0.14915	0.05876	17.01695	0.00876	114.0912
40	7.03978	0.14205	0.05828	17.15900	0.00828	120.7956

6.00% Compound Interest Factors

Periods	Single Payment		Uniform Series			
	Compound Amount Factor SCA 1	Present Worth Factor SPW 2	Capitol Recovery Factor UCR 3	Present Worth Factor UPW 4	Sinking Fund Factor USF 5	Compound Amount Factor UCA 6
1	1.06000	0.94340	1.06001	0.94339	1.00001	0.9999
2	1.12360	0.89000	0.54544	1.83337	0.48544	2.0599
3	1.19101	0.83962	0.37411	2.67298	0.31411	3.1835
4	1.26247	0.79210	0.28859	3.46508	0.22859	4.3745
5	1.33822	0.74726	0.23740	4.21233	0.17740	5.6370
6	1.41851	0.70496	0.20336	4.91728	0.14336	6.9752
7	1.50362	0.66506	0.17914	5.58234	0.11914	8.3937
8	1.59384	0.62741	0.16104	6.20975	0.10104	9.8973
9	1.68947	0.59190	0.14702	6.80165	0.08702	11.4911
10	1.79084	0.55840	0.13587	7.36004	0.07587	13.1806
11	1.89829	0.52679	0.12679	7.88683	0.06679	14.9714
12	2.01218	0.49697	0.11928	8.38380	0.05928	16.8697
13	2.13291	0.46884	0.11296	8.85263	0.05296	18.8818
14	2.26089	0.44230	0.10759	9.29493	0.04759	21.0147
15	2.39654	0.41727	0.10296	9.71220	0.04296	23.2756
16	2.54033	0.39365	0.09895	10.10585	0.03895	25.6721
17	2.69275	0.37137	0.09545	10.47721	0.03545	28.2124
18	2.85431	0.35035	0.09236	10.82755	0.03236	30.9052
19	3.02557	0.33052	0.08962	11.15807	0.02962	33.7594
20	3.20710	0.31181	0.08718	11.46987	0.02718	36.7850
21	3.39953	0.29416	0.08500	11.76403	0.02500	39.9921
22	3.60350	0.27751	0.08305	12.04153	0.02305	43.3916
23	3.81971	0.26180	0.08128	12.30333	0.02128	46.9950
24	4.04889	0.24698	0.07968	12.55031	0.01968	50.8147
25	4.29182	0.23300	0.07823	12.78331	0.01823	54.8636
26	4.54932	0.21981	0.07690	13.00312	0.01690	59.1553
27	4.82228	0.20737	0.07570	13.21049	0.01570	63.7046
28	5.11161	0.19563	0.07459	13.40613	0.01459	68.5269
29	5.41831	0.18456	0.07358	13.59068	0.01358	73.6384
30	5.74340	0.17411	0.07265	13.76479	0.01265	79.0567
31	6.08801	0.16426	0.07179	13.92905	0.01179	84.8000
32	6.45328	0.15496	0.07100	14.08401	0.01100	90.8880
33	6.84048	0.14619	0.07027	14.23020	0.01027	97.3412
34	7.25090	0.13791	0.06960	14.36810	0.00960	104.1816
35	7.68595	0.13011	0.06897	14.49821	0.00897	111.4325
36	8.14710	0.12274	0.06839	14.62096	0.00840	119.1183
37	8.63593	0.11580	0.06786	14.73675	0.00786	127.2654
38	9.15408	0.10924	0.06737	14.84599	0.00736	135.9013
39	9.70332	0.10306	0.06689	14.94905	0.00689	145.0553
40	10.28551	0.09722	0.06646	15.04628	0.00646	154.7585

7.00% Compound Interest Factors

Periods	Single Payment		Uniform Series			
	Compound Amount Factor SCA 1	Present Worth Factor SPW 2	Capitol Recovery Factor UCR 3	Present Worth Factor UPW 4	Sinking Fund Factor USF 5	Compound Amount Factor UCA 6
1	1.07000	0.93458	1.07000	0.93458	1.00000	1.0000
2	1.14490	0.87344	0.55310	1.80800	0.48310	2.0699
3	1.22504	0.81630	0.38105	2.62430	0.31105	3.2148
4	1.31079	0.76290	0.29523	3.38720	0.22523	4.4399
5	1.40255	0.71299	0.24389	4.10018	0.17389	5.7507
6	1.50073	0.66634	0.20980	4.76652	0.13980	7.1532
7	1.60578	0.62275	0.18555	5.38927	0.11555	8.6539
8	1.71818	0.58201	0.16747	5.97128	0.09747	10.2597
9	1.83845	0.54394	0.15349	6.51522	0.08349	11.9779
10	1.96714	0.50835	0.14238	7.02356	0.07238	13.8163
11	2.10485	0.47509	0.13336	7.49866	0.06336	15.7835
12	2.25218	0.44401	0.12590	7.94267	0.05590	17.8883
13	2.40984	0.41497	0.11965	8.35763	0.04965	20.1405
14	2.57852	0.38782	0.11435	8.74545	0.04435	22.5503
15	2.75902	0.36245	0.10979	9.10789	0.03979	25.1288
16	2.95215	0.33874	0.10586	9.44663	0.03586	27.8878
17	3.15880	0.31658	0.10243	9.76320	0.03243	30.8399
18	3.37991	0.29587	0.09941	10.05907	0.02941	33.9987
19	3.61651	0.27651	0.09675	10.33558	0.02675	37.3786
20	3.86966	0.25842	0.09439	10.59400	0.02439	40.9951
21	4.14054	0.24151	0.09229	10.83551	0.02229	44.8648
22	4.43037	0.22571	0.09041	11.06123	0.02041	49.0053
23	4.74050	0.21095	0.08871	11.27217	0.01871	53.4357
24	5.07233	0.19715	0.08719	11.46932	0.01719	58.1761
25	5.42739	0.18425	0.08581	11.65357	0.01581	63.2484
26	5.80731	0.17220	0.08456	11.82577	0.01456	68.6758
27	6.21382	0.16093	0.08343	11.98670	0.01343	74.4831
28	6.64878	0.15040	0.08239	12.13710	0.01239	80.6969
29	7.11420	0.14056	0.08145	12.27766	0.01145	87.3457
30	7.61219	0.13137	0.08059	12.40903	0.01059	94.4598
31	8.14504	0.12277	0.07980	12.53180	0.00980	102.0720
32	8.71519	0.11474	0.07907	12.64655	0.00907	110.2170
33	9.32525	0.10724	0.07841	12.75378	0.00841	118.9321
34	9.97802	0.10022	0.07780	12.85401	0.00780	128.2574
35	10.67647	0.09366	0.07723	12.94766	0.00723	138.2354
36	11.42382	0.08754	0.07672	13.03520	0.00672	148.9118
37	12.22349	0.08181	0.07624	13.11701	0.00624	160.3355
38	13.07913	0.07646	0.07580	13.19347	0.00580	172.5590
39	13.99466	0.07146	0.07539	13.26493	0.00539	185.6380
40	14.97429	0.06678	0.07501	13.33170	0.00501	199.6328

8.00% Compound Interest Factors

Periods	Single Payment		Uniform Series			
	Compound Amount Factor SCA	Present Worth Factor SPW	Capitol Recovery Factor UCR	Present Worth Factor UPW	Sinking Fund Factor USF	Compound Amount Factor UCA
	1	2	3	4	5	6
1	1.08000	0.92593	1.08000	0.92593	1.00000	1.0000
2	1.16640	0.85734	0.56077	1.78326	0.48077	2.0799
3	1.25971	0.79383	0.38803	2.57709	0.30803	3.2463
4	1.36049	0.73503	0.30192	3.31212	0.22192	4.5061
5	1.46933	0.68058	0.25046	3.99271	0.17046	5.8665
6	1.58687	0.63017	0.21632	4.62288	0.13623	7.3359
7	1.71382	0.58349	0.19207	5.20637	0.11207	8.9228
8	1.85093	0.54027	0.17401	5.74664	0.09401	10.6366
9	1.99900	0.50025	0.16008	6.24688	0.08008	12.4875
10	2.15892	0.46319	0.14903	6.71008	0.06903	14.4865
11	2.33164	0.42888	0.14008	7.13896	0.06008	16.6454
12	2.51817	0.39711	0.13270	7.53608	0.05270	18.9771
13	2.71962	0.36770	0.12652	7.90377	0.04652	21.4952
14	2.93719	0.34046	0.12130	8.24424	0.04130	24.2148
15	3.17216	0.31524	0.11683	8.55948	0.03683	27.1520
16	3.42594	0.29189	0.11298	8.85137	0.03298	30.3242
17	3.70001	0.27027	0.10963	9.12164	0.02963	33.7501
18	3.99601	0.25025	0.10670	9.37189	0.02670	37.4501
19	4.31569	0.23171	0.10413	9.60360	0.02413	41.4461
20	4.66095	0.21455	0.10185	9.81815	0.02185	45.7618
21	5.03383	0.19866	0.09983	10.01680	0.01983	50.4228
22	5.43653	0.18394	0.09803	10.20074	0.01803	55.4566
23	5.87145	0.17032	0.09642	10.37106	0.01642	60.8931
24	6.34117	0.15770	0.09498	10.52876	0.01498	66.7646
25	6.84846	0.14602	0.09368	10.67478	0.01368	73.1058
26	7.39634	0.13520	0.09251	10.80998	0.01251	79.9542
27	7.98805	0.12519	0.09145	10.93517	0.01145	87.3505
28	8.62709	0.11591	0.09049	11.05108	0.01049	95.3386
29	9.31726	0.10733	0.08962	11.15841	0.00962	103.9657
30	10.06264	0.09938	0.08883	11.25779	0.00883	113.2829
31	10.86765	0.09202	0.08811	11.34981	0.00811	123.3456
32	11.73706	0.08520	0.08745	11.43500	0.00745	134.2132
33	12.67602	0.07889	0.08685	11.51389	0.00685	145.9503
34	13.69010	0.07305	0.08630	11.58695	0.00630	158.6263
35	14.78531	0.06763	0.08580	11.65458	0.00580	172.3164
36	15.96813	0.06262	0.08534	11.71720	0.00534	187.1017
37	17.24557	0.05799	0.08492	11.77519	0.00492	203.0697
38	18.62521	0.05369	0.08454	11.82888	0.00454	220.3152
39	20.11523	0.04971	0.08419	11.87859	0.00419	238.9406
40	21.72446	0.04603	0.08386	11.92462	0.00386	259.0556

9.00% Compound Interest Factors

Periods	Single Payment		Uniform Series			
	Compound Amount Factor SCA 1	Present Worth Factor SPW 2	Capitol Recovery Factor UCR 3	Present Worth Factor UPW 4	Sinking Fund Factor USF 5	Compound Amount Factor UCA 6
1	1.09000	0.91743	1.09001	0.91742	1.00001	0.9999
2	1.18810	0.84168	0.56847	1.75910	0.47847	2.0899
3	1.29503	0.77219	0.39506	2.53128	0.30506	3.2780
4	1.41158	0.70843	0.30867	3.23970	0.21867	4.5730
5	1.53862	0.64993	0.25709	3.88963	0.16709	5.9846
6	1.67709	0.59627	0.22292	4.48589	0.13292	7.5232
7	1.82803	0.54704	0.19869	5.03292	0.10869	9.2003
8	1.99255	0.50187	0.18068	5.53479	0.09068	11.0283
9	2.17188	0.46043	0.16680	5.99522	0.07680	13.0208
10	2.36735	0.42241	0.15582	6.41763	0.06582	15.1927
11	2.58041	0.38754	0.14695	6.80516	0.05695	17.5600
12	2.81264	0.35554	0.13965	7.16070	0.04965	20.1404
13	3.06577	0.32618	0.13357	7.48687	0.04357	22.9530
14	3.34169	0.29925	0.12843	7.78612	0.03843	26.0188
15	3.64244	0.27454	0.12406	8.06066	0.03406	29.3604
16	3.97026	0.25187	0.12030	8.31253	0.03030	33.0028
17	4.32758	0.23108	0.11705	8.54361	0.02705	36.9731
18	4.71706	0.21200	0.11421	8.75560	0.02421	41.3006
19	5.14159	0.19449	0.11173	8.95009	0.02173	46.0176
20	5.60433	0.17843	0.10955	9.12852	0.01955	51.1592
21	6.10871	0.16370	0.10762	9.29222	0.01762	56.7634
22	6.65849	0.15018	0.10592	9.44240	0.01591	62.8721
23	7.25775	0.13778	0.10438	9.58019	0.01438	69.5305
24	7.91094	0.12641	0.10302	9.70659	0.01302	76.7882
25	8.62292	0.11597	0.10181	9.82256	0.01181	84.6991
26	9.39898	0.10639	0.10072	9.92896	0.01072	93.3220
27	10.24488	0.09761	0.09974	10.02656	0.00974	102.7209
28	11.16691	0.08955	0.09885	10.11612	0.00885	112.9657
29	12.17192	0.08216	0.09806	10.19827	0.00806	124.1325
30	13.26739	0.07537	0.09734	10.27364	0.00734	136.3043
31	14.46144	0.06915	0.09669	10.34279	0.00669	149.5716
32	15.76296	0.06344	0.09610	10.40624	0.00610	164.0329
33	17.18161	0.05820	0.09556	10.46444	0.00556	179.7957
34	18.72794	0.05340	0.09508	10.51783	0.00508	196.9772
35	20.41344	0.04899	0.09464	10.56681	0.00464	215.7050
36	22.25063	0.04494	0.09424	10.61176	0.00424	236.1181
37	24.25317	0.04123	0.09387	10.65299	0.00387	258.3686
38	26.43593	0.03783	0.09354	10.69082	0.00354	282.6213
39	28.81516	0.03470	0.09324	10.72552	0.00324	309.0573
40	31.40849	0.03184	0.09296	10.75736	0.00296	337.8720

10.00% Compound Interest Factors

Periods	Single Payment		Uniform Series			
	Compound Amount Factor SCA	Present Worth Factor SPW	Capitol Recovery Factor UCR	Present Worth Factor UPW	Sinking Fund Factor USF	Compound Amount Factor UCA
	1	2	3	4	5	6
1	1.10000	0.90909	1.10001	0.90909	1.00001	0.9999
2	1.21000	0.82645	0.57619	1.73552	0.47619	2.0999
3	1.33100	0.75132	0.40212	2.48684	0.30212	3.3099
4	1.46410	0.68302	0.31547	3.16985	0.21547	4.6409
5	1.61051	0.62092	0.26380	3.79077	0.16380	6.1050
6	1.77155	0.56448	0.22961	4.35524	0.12961	7.7155
7	1.94871	0.51316	0.20541	4.86840	0.10541	9.4870
8	2.14358	0.46651	0.18744	5.33490	0.08744	11.4358
9	2.35794	0.42410	0.17364	5.75900	0.07364	13.5793
10	2.59373	0.38555	0.16275	6.14455	0.06275	15.9372
11	2.85310	0.35050	0.15396	6.49504	0.05396	18.5309
12	3.13841	0.31863	0.14676	6.81367	0.04676	21.3840
13	3.45225	0.28967	0.14078	7.10334	0.04078	24.5224
14	3.79747	0.26333	0.13575	7.36667	0.03575	27.9746
15	4.17721	0.23939	0.13147	7.60606	0.03147	31.7721
16	4.59493	0.21763	0.12782	7.82369	0.02782	35.9493
17	5.05443	0.19785	0.12466	8.02153	0.02466	40.5442
18	5.55986	0.17986	0.12193	8.20139	0.02193	45.5986
19	6.11585	0.16351	0.11955	8.36491	0.01955	51.1584
20	6.72743	0.14865	0.11746	8.51355	0.01746	57.2742
21	7.40017	0.13513	0.11562	8.64868	0.01562	64.0016
22	8.14018	0.12285	0.11401	8.77152	0.01401	71.4017
23	8.95420	0.11168	0.11257	8.88321	0.01257	79.5419
24	9.84961	0.10153	0.11130	8.98473	0.01130	88.4960
25	10.83456	0.09230	0.11017	9.07703	0.01017	98.3456
26	11.91801	0.08391	0.10916	9.16094	0.00916	109.1801
27	13.10981	0.07628	0.10826	9.23722	0.00826	121.0980
28	14.42078	0.06934	0.10745	9.30655	0.00745	134.2078
29	15.86285	0.06304	0.10673	9.36959	0.00673	148.6285
30	17.44913	0.05731	0.10608	9.42691	0.00608	164.4912
31	19.19403	0.05210	0.10550	9.47901	0.00550	181.9402
32	21.11342	0.04736	0.10497	9.52637	0.00497	201.1341
33	23.22475	0.04306	0.10450	9.56943	0.00450	222.2475
34	25.54721	0.03914	0.10407	9.60857	0.00407	245.4722
35	28.10191	0.03558	0.10369	9.64416	0.00369	271.0190
36	30.91209	0.03235	0.10334	9.67650	0.00334	299.1208
37	34.00328	0.02941	0.10303	9.70591	0.00303	330.0327
38	37.40359	0.02674	0.10275	9.73265	0.00275	364.0356
39	41.14394	0.02430	0.10249	9.75695	0.00249	401.4392
40	45.25830	0.02210	0.10226	9.77905	0.00226	442.5827

Interest Tables

11.00% Compound Interest Factors

Periods	Single Payment		Uniform Series			
	Compound Amount Factor SCA 1	Present Worth Factor SPW 2	Capitol Recovery Factor UCR 3	Present Worth Factor UPW 4	Sinking Fund Factor USF 5	Compound Amount Factor UCA 6
1	1.11000	0.90090	1.11000	0.90090	1.00000	1.0000
2	1.23210	0.81162	0.58394	1.71251	0.47394	2.1099
3	1.36763	0.73119	0.40921	2.44370	0.29921	3.3420
4	1.51807	0.65873	0.32233	3.10244	0.21233	4.7097
5	1.68505	0.59345	0.27057	3.69589	0.16057	6.2277
6	1.87401	0.53464	0.23638	4.23053	0.12638	7.9128
7	2.07615	0.48166	0.21222	4.71218	0.10222	9.7832
8	2.30453	0.43393	0.19432	5.14611	0.08432	11.8593
9	2.55803	0.39093	0.18060	5.53704	0.07060	14.1639
10	2.83941	0.35219	0.16980	5.88922	0.05980	16.7219
11	3.15175	0.31728	0.16112	6.20651	0.05112	19.5613
12	3.49844	0.28584	0.15403	6.49235	0.04403	22.7130
13	3.88326	0.25752	0.14815	6.74986	0.03815	26.2114
14	4.31042	0.23200	0.14323	6.98186	0.03323	30.0947
15	4.78457	0.20901	0.13907	7.19086	0.02907	34.4051
16	5.31087	0.18829	0.13552	7.37915	0.02552	39.1897
17	5.89506	0.16963	0.13247	7.54879	0.02247	44.5005
18	6.54352	0.15282	0.12984	7.70161	0.01984	50.3956
19	7.26330	0.13768	0.12756	7.83929	0.01756	56.9390
20	8.06226	0.12403	0.12558	7.96332	0.01558	64.2023
21	8.94911	0.11174	0.12384	8.07506	0.01384	72.2646
22	9.93351	0.10067	0.12231	8.17574	0.01231	81.2136
23	11.02619	0.09069	0.12097	8.26643	0.01097	91.1471
24	12.23907	0.08171	0.11979	8.34813	0.00979	102.1733
25	13.58536	0.07361	0.11874	8.42174	0.00874	114.4123
26	15.07974	0.06631	0.11781	8.48806	0.00781	127.9976
27	16.73851	0.05974	0.11699	8.54780	0.00699	143.0773
28	18.57973	0.05382	0.11626	8.60162	0.00626	159.8157
29	20.62350	0.04849	0.11561	8.65011	0.00561	178.3954
30	22.89207	0.04368	0.11502	8.69379	0.00502	199.0188
31	25.41020	0.03935	0.11451	8.73314	0.00451	221.9109
32	28.20531	0.03545	0.11404	8.76860	0.00404	247.3211
33	31.30789	0.03194	0.11363	8.80054	0.00363	275.5261
34	34.75174	0.02878	0.11326	8.82931	0.00326	306.8337
35	38.57443	0.02592	0.11293	8.85524	0.00293	341.5856
36	42.81760	0.02335	0.11263	8.87860	0.00263	380.1599
37	47.52753	0.02104	0.11236	8.89963	0.00236	422.9772
38	52.75554	0.01896	0.11213	8.91859	0.00213	470.5046
39	58.55862	0.01708	0.11191	8.93567	0.00191	523.2600
40	65.00006	0.01538	0.11172	8.95105	0.00172	581.8186

12.00% Compound Interest Factors

Periods	Single Payment		Uniform Series			
	Compound Amount Factor SCA	Present Worth Factor SPW	Capitol Recovery Factor UCR	Present Worth Factor UPW	Sinking Fund Factor USF	Compound Amount Factor UCA
	1	2	3	4	5	6
1	1.12000	0.89286	1.12000	0.89286	1.00000	1.0000
2	1.25440	0.79719	0.59170	1.69005	0.47170	2.1199
3	1.40493	0.71178	0.41635	2.40183	0.29635	3.3743
4	1.57352	0.63552	0.32923	3.03734	0.20924	4.7793
5	1.76234	0.56743	0.27741	3.60477	0.15741	6.3528
6	1.97382	0.50633	0.24323	4.11140	0.12323	8.1151
7	2.21068	0.45235	0.21912	4.56375	0.09912	10.0890
8	2.47596	0.40388	0.20130	4.96764	0.08130	12.2996
9	2.77308	0.36061	0.18768	5.32825	0.06768	14.7756
10	3.10584	0.32197	0.17698	5.65022	0.05698	17.5486
11	3.47855	0.28748	0.16842	5.93770	0.04842	20.6545
12	3.89597	0.25668	0.16144	6.19437	0.04144	24.1330
13	4.36349	0.22917	0.15568	6.42355	0.03568	28.0290
14	4.88710	0.20462	0.15087	6.62817	0.03087	32.3925
15	5.47356	0.18270	0.14682	6.81086	0.02682	37.2796
16	6.13038	0.16312	0.14339	6.97398	0.02339	42.7531
17	6.86603	0.14564	0.14046	7.11963	0.02046	48.8835
18	7.68995	0.13004	0.13794	7.24967	0.01794	55.7495
19	8.61274	0.11611	0.13576	7.36578	0.01576	63.4395
20	9.64627	0.10367	0.13388	7.46944	0.01388	72.0522
21	10.80382	0.09256	0.13224	7.56201	0.01224	81.6985
22	12.10028	0.08264	0.13081	7.64465	0.01081	92.5023
23	13.55231	0.07379	0.12956	7.71844	0.00956	104.6026
24	15.17859	0.06588	0.12846	7.78432	0.00846	118.1549
25	17.00002	0.05882	0.12750	7.84314	0.00750	133.3334
26	19.04001	0.05252	0.12665	7.89566	0.00665	150.3334
27	21.32481	0.04689	0.12590	7.94255	0.00590	169.3734
28	23.88379	0.04187	0.12524	7.98442	0.00524	190.6982
29	26.74985	0.03738	0.12466	8.02181	0.00466	214.5821
30	29.95982	0.03338	0.12414	8.05518	0.00414	241.3319
31	33.55499	0.02980	0.12369	8.08499	0.00369	271.2915
32	37.58159	0.02661	0.12328	8.11160	0.00328	304.8464
33	42.09138	0.02376	0.12292	8.13535	0.00292	342.4279
34	47.14235	0.02121	0.12260	8.15657	0.00260	384.5195
35	52.79942	0.01894	0.12232	8.17550	0.00232	431.6616
36	59.13535	0.01691	0.12206	8.19242	0.00206	484.4611
37	66.23158	0.01510	0.12184	8.20751	0.00184	543.5964
38	74.17937	0.01348	0.12164	8.22099	0.00164	609.8278
39	83.08089	0.01204	0.12146	8.23303	0.00146	684.0073
40	93.05058	0.01075	0.12130	8.24378	0.00130	767.0881

13.00% Compound Interest Factors

Periods	Single Payment		Uniform Series			
	Compound Amount Factor SCA	Present Worth Factor SPW	Capitol Recovery Factor UCR	Present Worth Factor UPW	Sinking Fund Factor USF	Compound Amount Factor UCA
	1	2	3	4	5	6
1	1.13000	0.88496	1.13001	0.88495	1.00001	0.9999
2	1.27690	0.78315	0.59949	1.66809	0.46949	2.1299
3	1.44289	0.69305	0.42352	2.36114	0.29352	3.4068
4	1.63047	0.61332	0.33620	2.97446	0.20620	4.8497
5	1.84243	0.54276	0.28432	3.51722	0.15432	6.4802
6	2.08194	0.48032	0.25015	3.99753	0.12015	8.3226
7	2.35259	0.42506	0.22611	4.42259	0.09611	10.4045
8	2.65843	0.37616	0.20839	4.79875	0.07839	12.7571
9	3.00402	0.33289	0.19487	5.13164	0.06487	15.4155
10	3.39454	0.29459	0.18429	5.42623	0.05429	18.4195
11	3.83583	0.26070	0.17584	5.68692	0.04584	21.8140
12	4.33448	0.23071	0.16899	5.91763	0.03899	25.6498
13	4.89796	0.20417	0.16335	6.12180	0.03335	29.9843
14	5.53469	0.18068	0.15867	6.30247	0.02867	34.8822
15	6.25420	0.15989	0.15474	6.46237	0.02474	40.4169
16	7.06724	0.14150	0.15143	6.60386	0.02143	46.6710
17	7.98598	0.12522	0.14861	6.72908	0.01861	53.7382
18	9.02415	0.11081	0.14620	6.83990	0.01620	61.7242
19	10.19728	0.09807	0.14413	6.93796	0.01413	70.7482
20	11.52292	0.08678	0.14235	7.02475	0.01235	80.9455
21	13.02089	0.07680	0.14081	7.10154	0.01081	92.4683
22	14.71359	0.06796	0.13948	7.16951	0.00948	105.4891
23	16.62634	0.06015	0.13832	7.22965	0.00832	120.2026
24	18.78775	0.05323	0.13731	7.28288	0.00731	136.8288
25	21.23013	0.04710	0.13643	7.32998	0.00643	155.6164
26	23.99004	0.04168	0.13565	7.37166	0.00565	176.8464
27	27.10873	0.03689	0.13498	7.40855	0.00498	200.8364
28	30.63284	0.03264	0.13439	7.44120	0.00439	227.9451
29	34.61508	0.02889	0.13387	7.47009	0.00387	258.5773
30	39.11502	0.02557	0.13341	7.49565	0.00341	293.1923
31	44.19994	0.02262	0.13301	7.51828	0.00301	332.3071
32	49.94589	0.02002	0.13266	7.53830	0.00266	376.5068
33	56.43883	0.01772	0.13234	7.55601	0.00234	426.4523
34	63.77582	0.01568	0.13207	7.57169	0.00207	482.8908
35	72.06662	0.01388	0.13183	7.58557	0.00183	546.6662
36	81.43523	0.01228	0.13162	7.59785	0.00162	618.7324
37	92.02174	0.01087	0.13143	7.60872	0.00143	700.1672
38	103.98450	0.00962	0.13126	7.61833	0.00126	792.1884
39	117.50238	0.00851	0.13112	7.62685	0.00112	896.1721
40	132.77760	0.00753	0.13099	7.63438	0.00099	1013.6738

14.00% Compound Interest Factors

Periods	Single Payment		Uniform Series			
	Compound Amount Factor SCA 1	Present Worth Factor SPW 2	Capitol Recovery Factor UCR 3	Present Worth Factor UPW 4	Sinking Fund Factor USF 5	Compound Amount Factor UCA 6
1	1.14000	0.87719	1.14000	0.87719	1.00000	1.0000
2	1.29960	0.76947	0.60729	1.64665	0.46729	2.1399
3	1.48154	0.67497	0.43073	2.32162	0.29073	3.4395
4	1.68896	0.59208	0.34321	2.91370	0.20321	4.9211
5	1.92541	0.51937	0.29128	3.43307	0.15128	6.6100
6	2.19497	0.45559	0.25716	3.88866	0.11716	8.5354
7	2.50226	0.39964	0.23319	4.28829	0.09319	10.7304
8	2.85257	0.35056	0.21557	4.63885	0.07557	13.2326
9	3.25193	0.30751	0.20217	4.94636	0.06217	16.0852
10	3.70720	0.26975	0.19171	5.21611	0.05171	19.3371
11	4.22621	0.23662	0.18339	5.45272	0.04339	23.0443
12	4.81787	0.20756	0.17667	5.66028	0.03667	27.2705
13	5.49327	0.18207	0.17116	5.84325	0.03116	32.0883
14	6.26130	0.15971	0.16661	6.00206	0.02661	37.5807
15	7.13788	0.14010	0.16281	6.14216	0.02281	43.8420
16	8.13718	0.12289	0.15962	6.26505	0.01962	50.9798
17	9.27638	0.10780	0.15692	6.37285	0.01692	59.1170
18	10.57507	0.09456	0.15462	6.46742	0.01462	68.3933
19	12.05557	0.08295	0.15266	6.55037	0.01266	78.9683
20	13.74334	0.07276	0.15099	6.62313	0.01099	91.0238
21	15.66740	0.06383	0.14954	6.68695	0.00954	104.7672
22	17.86082	0.05599	0.14830	6.74294	0.00830	120.4344
23	20.36133	0.04911	0.14723	6.79206	0.00723	138.2952
24	23.21190	0.04308	0.14630	6.83514	0.00630	158.6564
25	26.46155	0.03779	0.14550	6.87293	0.00550	181.8683
26	30.16615	0.03315	0.14480	6.90608	0.00480	208.3298
27	34.38940	0.02908	0.14419	6.93515	0.00419	238.4958
28	39.20390	0.02551	0.14366	6.96066	0.00366	272.8850
29	44.69241	0.02238	0.14320	6.98304	0.00320	312.0886
30	50.94933	0.01963	0.14280	7.00266	0.00280	356.7807
31	58.08221	0.01722	0.14245	7.01988	0.00245	407.7299
32	66.21368	0.01510	0.14215	7.03498	0.00215	465.8120
33	75.48357	0.01325	0.14188	7.04823	0.00188	532.0253
34	86.05121	0.01162	0.14165	7.05985	0.00165	607.5085
35	98.09833	0.01019	0.14144	7.07005	0.00144	693.5595
36	111.83205	0.00894	0.14126	7.07899	0.00126	791.6574
37	127.48846	0.00784	0.14111	7.08683	0.00111	903.4890
38	145.33676	0.00688	0.14097	7.09371	0.00097	1030.9770
39	165.68382	0.00604	0.14085	7.09975	0.00085	1176.3132
40	188.87946	0.00529	0.14075	7.10504	0.00075	1341.9963

15.00% Compound Interest Factors

Periods	Single Payment		Uniform Series			
	Compound Amount Factor SCA 1	Present Worth Factor SPW 2	Capitol Recovery Factor UCR 3	Present Worth Factor UPW 4	Sinking Fund Factor USF 5	Compound Amount Factor UCA 6
1	1.15000	0.86957	1.15000	0.86956	1.00000	1.0000
2	1.32250	0.75614	0.61512	1.62570	0.46512	2.1499
3	1.52087	0.65752	0.43798	2.28322	0.28798	3.4724
4	1.74900	0.57175	0.35027	2.85497	0.20027	4.9933
5	2.01135	0.49718	0.29832	3.35215	0.14832	6.7423
6	2.31306	0.43233	0.26424	3.78448	0.11424	8.7537
7	2.66001	0.37594	0.24036	4.16041	0.09036	11.0667
8	3.05901	0.32690	0.22285	4.48732	0.07285	13.7267
9	3.51787	0.28426	0.20957	4.77158	0.05957	16.7857
10	4.04554	0.24719	0.19925	5.01876	0.04925	20.3036
11	4.65237	0.21494	0.19107	5.23371	0.04107	24.3491
12	5.35023	0.18691	0.18448	5.42062	0.03448	29.0015
13	6.15276	0.16253	0.17911	5.58314	0.02911	34.3517
14	7.07567	0.14133	0.17469	5.72448	0.02469	40.5044
15	8.13702	0.12290	0.17102	5.84737	0.02102	47.5801
16	9.35757	0.10687	0.16795	5.95423	0.01795	55.7171
17	10.76120	0.09293	0.16537	6.04716	0.01537	65.0747
18	12.37538	0.08081	0.16319	6.12797	0.01319	75.8358
19	14.23168	0.07027	0.16134	6.19823	0.01134	88.2112
20	16.36642	0.06110	0.15976	6.25933	0.00976	102.4428
21	18.82138	0.05313	0.15842	6.31246	0.00842	118.8092
22	21.64458	0.04620	0.15727	6.35866	0.00727	137.6305
23	24.89127	0.04017	0.15628	6.39884	0.00628	159.2752
24	28.62494	0.03493	0.15543	6.43377	0.00543	184.1662
25	32.91867	0.03038	0.15470	6.46415	0.00470	212.7911
26	37.85646	0.02642	0.15407	6.49057	0.00407	245.7099
27	43.53491	0.02297	0.15353	6.51354	0.00353	283.5659
28	50.06514	0.01997	0.15306	6.53351	0.00306	327.1008
29	57.57489	0.01737	0.15265	6.55088	0.00265	377.1657
30	66.21111	0.01510	0.15230	6.56598	0.00230	434.7407
31	76.14275	0.01313	0.15200	6.57911	0.00200	500.9516
32	87.56413	0.01142	0.15173	6.59054	0.00173	577.0942
33	100.69872	0.00993	0.15150	6.60047	0.00150	664.6582
34	115.80348	0.00864	0.15131	6.60910	0.00131	765.3566
35	133.17397	0.00751	0.15113	6.61661	0.00113	881.1599
36	153.15001	0.00653	0.15099	6.62314	0.00099	1014.3335
37	176.12247	0.00568	0.15086	6.62882	0.00086	1167.4834
38	202.54076	0.00494	0.15074	6.63375	0.00074	1343.6052
39	232.92180	0.00429	0.15065	6.63805	0.00065	1546.1457
40	267.85986	0.00373	0.15056	6.64178	0.00056	1779.0664

16.00% Compound Interest Factors

Periods	Single Payment		Uniform Series			
	Compound Amount Factor SCA 1	Present Worth Factor SPW 2	Capitol Recovery Factor UCR 3	Present Worth Factor UPW 4	Sinking Fund Factor USF 5	Compound Amount Factor UCA 6
1	1.16000	0.86207	1.16000	0.86207	1.00000	1.0000
2	1.34560	0.74316	0.62296	1.60523	0.46296	2.1599
3	1.56089	0.64066	0.44526	2.24589	0.28526	3.5055
4	1.81064	0.55229	0.35738	2.79818	0.19738	5.0664
5	2.10034	0.47611	0.30541	3.27429	0.14541	6.8771
6	2.43639	0.41044	0.27139	3.68473	0.11139	8.9774
7	2.82622	0.35383	0.24761	4.03856	0.08761	11.4138
8	3.27841	0.30503	0.23022	4.34359	0.07022	14.2400
9	3.80296	0.26295	0.21708	4.60654	0.05708	17.5184
10	4.41143	0.22668	0.20690	4.83322	0.04690	21.3214
11	5.11726	0.19542	0.19886	5.02864	0.03886	25.7328
12	5.93602	0.16846	0.19241	5.19710	0.03241	30.8501
13	6.88578	0.14523	0.18718	5.34233	0.02718	36.7861
14	7.98750	0.12520	0.18920	5.46753	0.02290	43.6718
15	9.26550	0.10793	0.17936	5.57546	0.01936	51.6593
16	10.74798	0.09304	0.17641	5.66850	0.01641	60.9248
17	12.46766	0.08021	0.17395	5.74870	0.01395	71.6728
18	14.46248	0.06914	0.17188	5.81785	0.01188	84.1404
19	16.77647	0.05961	0.17014	5.87746	0.01014	98.6029
20	19.46069	0.05139	0.16867	5.92884	0.00867	115.3793
21	22.57442	0.04430	0.16742	5.97314	0.00742	134.8401
22	26.18631	0.03819	0.16635	6.01133	0.00635	157.4144
23	30.37613	0.03292	0.16545	6.04425	0.00545	183.6008
24	35.23630	0.02838	0.16467	6.07263	0.00467	213.9768
25	40.87410	0.02447	0.16401	6.09709	0.00401	249.2131
26	47.41396	0.02109	0.16345	6.11818	0.00345	290.0871
27	55.00018	0.01818	0.16296	6.13636	0.00296	337.5009
28	63.80020	0.01567	0.16255	6.15204	0.00255	392.5009
29	74.00822	0.01351	0.16219	6.16555	0.00219	456.3012
30	85.84953	0.01165	0.16189	6.17720	0.00189	530.3093
31	99.58545	0.01004	0.16162	6.18724	0.00162	616.1589
32	115.51910	0.00866	0.16140	6.19590	0.00140	715.7441
33	134.00214	0.00746	0.16120	6.20336	0.00120	831.2631
34	155.44246	0.00643	0.16104	6.20979	0.00104	965.2651
35	180.31323	0.00555	0.16089	6.21534	0.00089	1120.7075
36	209.16333	0.00478	0.16077	6.22012	0.00077	1301.0205
37	242.62943	0.00412	0.16066	6.22424	0.00066	1510.1835
38	281.44995	0.00355	0.16057	6.22779	0.00057	1752.8122
39	326.48193	0.00306	0.16049	6.23086	0.00049	2034.2624
40	378.71899	0.00264	0.16042	6.23350	0.00042	2360.7436

17.00% Compound Interest Factors

Periods	Single Payment		Uniform Series			
	Compound Amount Factor SCA 1	Present Worth Factor SPW 2	Capitol Recovery Factor UCR 3	Present Worth Factor UPW 4	Sinking Fund Factor USF 5	Compound Amount Factor UCA 6
1	1.17000	0.85470	1.17000	0.85470	1.00000	1.0000
2	1.36890	0.73051	0.63083	1.58521	0.46083	2.1699
3	1.60161	0.62437	0.45258	2.20958	0.28258	3.5388
4	1.87388	0.53365	0.36453	2.74323	0.19453	5.1404
5	2.19244	0.45611	0.31256	3.19934	0.14256	7.0143
6	2.56515	0.38984	0.27862	3.58917	0.10862	9.2067
7	3.00123	0.33320	0.25495	3.93327	0.08495	11.7719
8	3.51143	0.28478	0.23769	4.20715	0.06769	14.7731
9	4.10837	0.24341	0.22469	4.45056	0.05469	18.2845
10	4.80679	0.20804	0.21466	4.65859	0.04466	22.3928
11	5.62394	0.17781	0.20677	4.83641	0.03677	27.1996
12	6.58001	0.15198	0.20047	4.98838	0.03047	32.8235
13	7.69860	0.12989	0.19538	5.11827	0.02538	39.4035
14	9.00736	0.11102	0.19123	5.22929	0.02123	47.1021
15	10.53860	0.09489	0.18782	5.32418	0.01782	56.1094
16	12.33016	0.08110	0.18500	5.40528	0.01500	66.6479
17	14.42627	0.06932	0.18266	5.47460	0.01266	78.9780
18	16.87872	0.05925	0.18071	5.53385	0.01071	93.4042
19	19.74809	0.05064	0.17907	5.58448	0.00907	110.2829
20	23.10524	0.04328	0.17769	5.62776	0.00769	130.0308
21	27.03311	0.03699	0.17653	5.66475	0.00653	153.1359
22	31.62872	0.03162	0.17555	5.69637	0.00555	180.1690
23	37.00558	0.02702	0.17472	5.72339	0.00472	211.7975
24	43.29649	0.02310	0.17402	5.74649	0.00402	248.8029
25	50.65688	0.01974	0.17342	5.76623	0.00342	292.0991
26	59.26849	0.01687	0.17292	5.78310	0.00292	342.7556
27	69.34409	0.01442	0.17249	5.79753	0.00249	402.0239
28	81.13252	0.01233	0.17212	5.80985	0.00212	471.3676
29	94.92497	0.01053	0.17181	5.82039	0.00181	552.4997
30	111.06215	0.00900	0.17154	5.82939	0.00154	647.4243
31	129.94261	0.00770	0.17132	5.83709	0.00132	758.4858
32	152.03275	0.00658	0.17113	5.84366	0.00113	888.4277
33	177.87817	0.00562	0.17096	5.84928	0.00096	1040.4597
34	208.11731	0.00480	0.17082	5.85409	0.00082	1218.3369
35	243.49707	0.00411	0.17070	5.85820	0.00070	1426.4533
36	284.89136	0.00351	0.17060	5.86170	0.00060	1669.9489
37	333.32251	0.00300	0.17051	5.86471	0.00051	1954.8381
38	389.98706	0.00256	0.17044	5.86727	0.00044	2288.1596
39	456.28467	0.00219	0.17037	5.86946	0.00037	2678.1462
40	533.85254	0.00187	0.17032	5.87133	0.00032	3134.4267

18.00% Compound Interest Factors

Periods	Single Payment		Uniform Series			
	Compound Amount Factor SCA 1	Present Worth Factor SPW 2	Capitol Recovery Factor UCR 3	Present Worth Factor UPW 4	Sinking Fund Factor USF 5	Compound Amount Factor UCA 6
1	1.18000	0.84746	1.18000	0.84746	1.00000	1.0000
2	1.39240	0.71819	0.63872	1.56564	0.45872	2.1799
3	1.64303	0.60863	0.45993	2.17427	0.27993	3.5723
4	1.93877	0.51579	0.37174	2.69005	0.19174	5.2154
5	2.28775	0.43711	0.31978	3.12716	0.13978	7.1541
6	2.69954	0.37043	0.28591	3.49760	0.10591	9.4419
7	3.18546	0.31393	0.26236	3.81152	0.08236	12.1414
8	3.75884	0.26604	0.24524	4.07756	0.06524	15.3269
9	4.43543	0.22546	0.23240	4.30301	0.05240	19.0857
10	5.23381	0.19107	0.22251	4.49408	0.04251	23.5211
11	6.17589	0.16192	0.21478	4.65600	0.03478	28.7549
12	7.28754	0.13722	0.20863	4.79322	0.02863	34.9308
13	8.59930	0.11629	0.20369	4.90951	0.02369	42.2183
14	10.14717	0.09855	0.19968	5.00806	0.01968	50.8175
15	11.97365	0.08352	0.19640	5.09158	0.01640	60.9647
16	14.12890	0.07078	0.19371	5.16235	0.01371	72.9383
17	16.67209	0.05998	0.19149	5.22233	0.01149	87.0671
18	19.67305	0.05083	0.18964	5.27316	0.00964	103.7391
19	23.21419	0.04308	0.18810	5.31624	0.00810	123.4121
20	27.39273	0.03651	0.18682	5.35275	0.00682	146.6263
21	32.32339	0.03094	0.18575	5.38368	0.00575	174.0188
22	38.14159	0.02622	0.18485	5.40990	0.00485	206.3421
23	45.00705	0.02222	0.18409	5.43212	0.00409	244.4836
24	53.10829	0.01883	0.18345	5.45095	0.00345	289.4904
25	62.66776	0.01596	0.18292	5.46690	0.00292	342.5986
26	73.94792	0.01352	0.18247	5.48043	0.00247	405.2661
27	87.25850	0.01146	0.18209	5.49189	0.00209	479.2138
28	102.96497	0.00971	0.18177	5.50160	0.00177	566.4719
29	121.49860	0.00823	0.18149	5.50983	0.00149	669.4365
30	143.36827	0.00698	0.18126	5.51681	0.00126	790.9348
31	169.17447	0.00591	0.18107	5.52272	0.00107	934.3027
32	199.62576	0.00501	0.18091	5.52773	0.00091	1103.4765
33	235.55827	0.00425	0.18077	5.53197	0.00077	1303.1015
34	277.95850	0.00360	0.18065	5.53557	0.00065	1538.6586
35	327.99097	0.00305	0.18055	5.53862	0.00055	1816.6167
36	387.02905	0.00258	0.18047	5.54120	0.00047	2144.6064
37	456.69409	0.00219	0.18039	5.54339	0.00040	2531.6335
38	538.89868	0.00186	0.18033	5.54525	0.00033	2988.3271
39	635.90015	0.00157	0.18028	5.54682	0.00028	3527.2243
40	750.36182	0.00133	0.18024	5.54815	0.00024	4163.1210

19.00% Compound Interest Factors

Periods	Single Payment		Uniform Series			
	Compound Amount Factor SCA 1	Present Worth Factor SPW 2	Capitol Recovery Factor UCR 3	Present Worth Factor UPW 4	Sinking Fund Factor USF 5	Compound Amount Factor UCA 6
1	1.19000	0.84034	1.19000	0.84034	1.00000	1.0000
2	1.41610	0.70617	0.64662	1.54650	0.45662	2.1899
3	1.68516	0.59342	0.46731	2.13991	0.27731	3.6060
4	2.00534	0.49867	0.37899	2.63858	0.18899	5.2912
5	2.38635	0.41095	0.32705	3.05763	0.13705	7.2965
6	2.83975	0.35214	0.29327	3.40977	0.10327	9.6829
7	3.37931	0.29592	0.26986	3.70569	0.07986	12.5226
8	4.02137	0.24867	0.25289	3.95436	0.06289	15.9019
9	4.78543	0.20897	0.24019	4.16333	0.05019	19.9233
10	5.69466	0.17560	0.23047	4.33893	0.04047	24.7087
11	6.77665	0.14757	0.22289	4.48650	0.03289	30.4034
12	8.06421	0.12400	0.21690	4.61050	0.02690	37.1800
13	9.59640	0.10421	0.21210	4.71471	0.02210	45.2442
14	11.41972	0.08757	0.20823	4.80228	0.01823	54.8406
15	13.58946	0.07359	0.20509	4.87586	0.01509	66.2603
16	16.17145	0.06184	0.20252	4.93770	0.01252	79.8497
17	19.24400	0.05196	0.20041	4.98966	0.01041	96.0210
18	22.90036	0.04367	0.19868	5.03333	0.00868	115.2650
19	27.25142	0.03670	0.19724	5.07003	0.00724	138.1653
20	32.42918	0.03084	0.19605	5.10086	0.00605	165.4168
21	38.59071	0.02591	0.19505	5.12677	0.00505	197.8459
22	45.92294	0.02178	0.19423	5.14855	0.00423	236.4365
23	54.64828	0.01830	0.19354	5.16685	0.00354	282.3593
24	65.03143	0.01538	0.19297	5.18223	0.00297	337.0073
25	77.38739	0.01292	0.19249	5.19515	0.00249	402.0388
26	92.09096	0.01086	0.19209	5.20601	0.00209	479.4260
27	109.58820	0.00913	0.19175	5.21513	0.00175	571.5166
28	130.40991	0.00767	0.19147	5.22280	0.00147	681.1047
29	155.18774	0.00644	0.19123	5.22925	0.00123	811.5144
30	184.67336	0.00541	0.19103	5.23466	0.00103	966.7019
31	219.76122	0.00455	0.19087	5.23921	0.00087	1151.3750
32	261.51562	0.00382	0.19073	5.24303	0.00073	1371.1350
33	311.20361	0.00321	0.19061	5.24625	0.00061	1632.6506
34	370.33203	0.00270	0.19051	5.24895	0.00051	1943.8532
35	440.69507	0.00227	0.19043	5.25122	0.00043	2314.1845
36	524.42700	0.00191	0.19036	5.25312	0.00036	2754.8793
37	624.06787	0.00160	0.19030	5.25472	0.00030	3279.3068
38	742.64062	0.00135	0.19026	5.25607	0.00026	3903.3730
39	883.74194	0.00113	0.19022	5.25720	0.00022	4646.0078
40	1051.65259	0.00095	0.19018	5.25815	0.00018	5529.7500

20.00% Compound Interest Factors

Periods	Single Payment		Uniform Series			
	Compound Amount Factor SCA	Present Worth Factor SPW	Capitol Recovery Factor UCR	Present Worth Factor UPW	Sinking Fund Factor USF	Compound Amount Factor UCA
	1	2	3	4	5	6
1	1.20000	0.83333	1.20000	0.83333	1.00000	1.0000
2	1.44000	0.69445	0.65455	1.52777	0.45455	2.1999
3	1.72800	0.57870	0.47473	2.10648	0.27473	3.6399
4	2.07360	0.48225	0.38629	2.58873	0.18629	5.3679
5	2.48832	0.40188	0.33438	2.99061	0.13438	7.4415
6	2.98598	0.33490	0.30071	3.32551	0.10071	9.9299
7	3.58318	0.27908	0.27742	3.60459	0.07742	12.9158
8	4.29981	0.23257	0.26061	3.83716	0.06061	16.4990
9	5.15977	0.19381	0.24808	4.03097	0.04808	20.7988
10	6.19173	0.16151	0.23852	4.19247	0.03852	25.9586
11	7.43007	0.13459	0.23110	4.32706	0.03110	32.1503
12	8.91608	0.11216	0.22526	4.43922	0.02527	39.5804
13	10.69930	0.09346	0.22062	4.53268	0.02062	48.4964
14	12.83916	0.07789	0.21689	4.61057	0.01689	59.1957
15	15.40698	0.06491	0.21388	4.67547	0.01388	72.0349
16	18.48837	0.05409	0.21144	4.72956	0.01144	87.4418
17	22.18605	0.04507	0.20944	4.77463	0.00944	105.9302
18	26.62325	0.03756	0.20781	4.81220	0.00781	128.1162
19	31.94789	0.03130	0.20646	4.84350	0.00646	154.7395
20	38.33746	0.02608	0.20536	4.86958	0.00536	186.6874
21	46.00496	0.02174	0.20444	4.89132	0.00444	225.0249
22	55.20595	0.01811	0.20369	4.90943	0.00369	271.0295
23	66.24712	0.01509	0.20307	4.92453	0.00307	326.2356
24	79.49654	0.01258	0.20255	4.93710	0.00255	392.4826
25	95.39583	0.01048	0.20212	4.94759	0.00212	471.9790
26	114.47498	0.00874	0.20176	4.95632	0.00176	567.3747
27	137.36995	0.00728	0.20147	4.96360	0.00147	681.8496
28	164.84392	0.00607	0.20122	4.96967	0.00122	819.2194
29	197.81267	0.00506	0.20102	4.97472	0.00102	984.0634
30	237.37517	0.00421	0.20085	4.97894	0.00085	1181.8759
31	284.85010	0.00351	0.20070	4.98245	0.00070	1419.2509
32	341.82007	0.00293	0.20059	4.98537	0.00059	1704.1003
33	410.18408	0.00244	0.20049	4.98781	0.00049	2045.9209
34	492.22070	0.00203	0.20041	4.98984	0.00041	2456.1040
35	590.66479	0.00169	0.20034	4.99154	0.00034	2948.3254
36	708.79761	0.00141	0.20028	4.99295	0.00028	3538.9895
37	850.55713	0.00118	0.20024	4.99412	0.00024	4247.7851
38	1020.66846	0.00098	0.20020	4.99510	0.00020	5098.3398
39	1224.80200	0.00082	0.20016	4.99592	0.00016	6119.0078
40	1469.76221	0.00068	0.20014	4.99660	0.00014	7343.8085

21.00% Compound Interest Factors

Periods	Single Payment		Uniform Series			
	Compound Amount Factor SCA 1	Present Worth Factor SPW 2	Capitol Recovery Factor UCR 3	Present Worth Factor UPW 4	Sinking Fund Factor USF 5	Compound Amount Factor UCA 6
1	1.21000	0.82645	1.21000	0.82644	1.00000	1.0000
2	1.46410	0.68301	0.66249	1.50945	0.45249	2.2099
3	1.77156	0.56448	0.48218	2.07393	0.27218	3.6740
4	2.14358	0.46651	0.39363	2.54043	0.18363	5.4456
5	2.59373	0.38554	0.34177	2.92598	0.13177	7.5892
6	3.13841	0.31863	0.30820	3.24461	0.09820	10.1829
7	3.79748	0.26333	0.28507	3.50794	0.07507	13.3213
8	4.59494	0.21763	0.26842	3.72557	0.05842	17.1187
9	5.55988	0.17986	0.25605	3.90543	0.04605	21.7137
10	6.72745	0.14864	0.24667	4.05407	0.03667	27.2735
11	8.14021	0.12285	0.23941	4.17692	0.02941	34.0009
12	9.84964	0.10153	0.23373	4.27845	0.02373	42.1411
13	11.91806	0.08391	0.22923	4.36235	0.01923	51.9907
14	14.42084	0.06934	0.22565	4.43170	0.01565	63.9087
15	17.44920	0.05731	0.22277	4.48901	0.01277	78.3295
16	21.11351	0.04736	0.22044	4.53637	0.01044	95.7786
17	25.54733	0.03914	0.21855	4.57551	0.00855	116.8920
18	30.91225	0.03235	0.21702	4.60786	0.00702	142.4393
19	37.40379	0.02674	0.21577	4.63459	0.00577	173.3514
20	45.25856	0.02210	0.21474	4.65669	0.00474	210.7551
21	54.76282	0.01826	0.21391	4.67495	0.00391	256.0134
22	66.26297	0.01509	0.21322	4.69004	0.00322	310.7758
23	80.17813	0.01247	0.21265	4.70251	0.00265	377.0385
24	97.01546	0.01031	0.21219	4.71282	0.00219	457.2163
25	117.38863	0.00852	0.21180	4.72134	0.00180	554.2314
26	142.04013	0.00704	0.21149	4.72838	0.00149	671.6196
27	171.86842	0.00582	0.21123	4.73420	0.00123	813.6591
28	207.96065	0.00481	0.21101	4.73901	0.00101	985.5268
29	251.63219	0.00397	0.21084	4.74298	0.00084	1193.4865
30	304.47461	0.00328	0.21069	4.74627	0.00069	1445.1174
31	368.41406	0.00271	0.21057	4.74898	0.00057	1749.5913
32	445.78076	0.00224	0.21047	4.75122	0.00047	2118.0041
33	539.39429	0.00185	0.21039	4.75308	0.00039	2563.7829
34	652.66650	0.00153	0.21032	4.75461	0.00032	3103.1743
35	789.72583	0.00127	0.21027	4.75588	0.00027	3755.8391
36	955.56763	0.00105	0.21022	4.75692	0.00022	4545.5585
37	1156.23608	0.00086	0.21018	4.75779	0.00018	5501.1250
38	1399.04468	0.00071	0.21015	4.75850	0.00015	6657.3554
39	1692.84277	0.00059	0.21012	4.75909	0.00012	8056.3945
40	2048.33813	0.00049	0.21010	4.75958	0.00010	9749.2304

22.00% Compound Interest Factors

	Single Payment		Uniform Series			
Periods	Compound Amount Factor SCA 1	Present Worth Factor SPW 2	Capitol Recovery Factor UCR 3	Present Worth Factor UPW 4	Sinking Fund Factor USF 5	Compound Amount Factor UCA 6
1	1.22000	0.81967	1.22000	0.81967	1.00000	1.0000
2	1.48840	0.67186	0.67045	1.49153	0.45045	2.2199
3	1.81584	0.55071	0.48966	2.04224	0.26966	3.7083
4	2.21533	0.45140	0.40102	2.49364	0.18102	5.5242
5	2.70270	0.37000	0.34921	2.86363	0.12921	7.7395
6	3.29729	0.30328	0.31576	3.16691	0.09576	10.4422
7	4.02269	0.24859	0.29278	3.41550	0.07278	13.7395
8	4.90768	0.20376	0.27630	3.61926	0.05630	17.7621
9	5.98737	0.16702	0.26411	3.78628	0.04411	22.6698
10	7.30459	0.13690	0.25490	3.92318	0.03490	28.6572
11	8.91159	0.11221	0.24781	4.03539	0.02781	35.9617
12	10.87214	0.09198	0.24228	4.12737	0.02228	44.8733
13	13.26400	0.07539	0.23794	4.20276	0.01794	55.7454
14	16.18207	0.06180	0.23449	4.26456	0.01449	69.0094
15	19.74211	0.05065	0.23174	4.31521	0.01174	85.1914
16	24.08537	0.04152	0.22953	4.35673	0.00953	104.9335
17	29.38412	0.03403	0.22775	4.39076	0.00775	129.0187
18	35.84862	0.02790	0.22631	4.41866	0.00631	158.4028
19	43.73529	0.02286	0.22515	4.44152	0.00515	194.2514
20	53.35703	0.01874	0.22420	4.46026	0.00420	237.9865
21	65.09555	0.01536	0.22343	4.47563	0.00343	291.3432
22	79.41652	0.01259	0.22281	4.48822	0.00281	356.4384
23	96.88811	0.01032	0.22229	4.49854	0.00229	435.8549
24	118.20341	0.00846	0.22188	4.50700	0.00188	532.7426
25	144.20810	0.00693	0.22154	4.51393	0.00154	650.9458
26	175.93378	0.00568	0.22126	4.51962	0.00126	795.1533
27	214.63908	0.00466	0.22103	4.52428	0.00103	971.0866
28	261.85937	0.00382	0.22084	4.52810	0.00084	1185.7241
29	319.46826	0.00313	0.22069	4.53123	0.00069	1447.5830
30	389.75122	0.00257	0.22057	4.53379	0.00057	1767.0507
31	475.59609	0.00210	0.22046	4.53590	0.00046	2156.8010
32	580.10498	0.00172	0.22038	4.53762	0.00038	2632.2959
33	707.72778	0.00141	0.22031	4.53903	0.00031	3212.3984
34	863.42749	0.00116	0.22026	4.54019	0.00026	3920.1247
35	1053.38086	0.00095	0.22021	4.54114	0.00021	4783.5468
36	1285.12402	0.00078	0.22017	4.54192	0.00017	5836.9257
37	1567.85059	0.00064	0.22014	4.54256	0.00014	7122.0468
38	1912.77661	0.00052	0.22012	4.54308	0.00012	8689.8906
39	2333.58618	0.00043	0.22009	4.54351	0.00009	10602.6602
40	2846.97363	0.00035	0.22008	4.54386	0.00008	12936.2422

23.00% Compound Interest Factors

Periods	Single Payment		Uniform Series			
	Compound Amount Factor SCA 1	Present Worth Factor SPW 2	Capitol Recovery Factor UCR 3	Present Worth Factor UPW 4	Sinking Fund Factor USF 5	Compound Amount Factor UCA 6
1	1.23000	0.81301	1.23000	0.81301	1.00000	1.0000
2	1.51290	0.66098	0.67843	1.47399	0.44843	2.2299
3	1.86086	0.53738	0.49717	2.01137	0.26717	3.7428
4	2.28886	0.43690	0.40845	2.44827	0.17845	5.6037
5	2.81530	0.35520	0.35670	2.80347	0.12670	7.8926
6	3.46282	0.28878	0.32339	3.09225	0.09339	10.7079
7	4.25926	0.23478	0.30057	3.32703	0.07057	14.1707
8	5.23889	0.19088	0.28426	3.51791	0.05426	18.4299
9	6.44384	0.15519	0.27225	3.67310	0.04225	23.6688
10	7.92592	0.12617	0.26321	3.79927	0.03321	30.1126
11	9.74887	0.10258	0.25629	3.90184	0.02629	38.0385
12	11.99111	0.08340	0.25093	3.98524	0.02093	47.7874
13	14.74906	0.06780	0.24673	4.05304	0.01673	59.7785
14	18.14133	0.05512	0.24342	4.10816	0.01342	74.5275
15	22.31383	0.04482	0.24079	4.15298	0.01079	92.6688
16	27.44600	0.03644	0.23870	4.18941	0.00870	114.9826
17	33.75858	0.02962	0.23702	4.21903	0.00702	142.4285
18	41.52303	0.02408	0.23568	4.24312	0.00568	176.1870
19	51.07330	0.01958	0.23459	4.26270	0.00459	217.7100
20	62.82014	0.01592	0.23372	4.27861	0.00372	268.7832
21	77.26875	0.01294	0.23302	4.29156	0.00302	331.6030
22	95.04053	0.01052	0.23245	4.30208	0.00245	408.3718
23	116.89981	0.00855	0.23198	4.31063	0.00198	503.9121
24	143.78671	0.00695	0.23161	4.31759	0.00161	620.8117
25	176.85760	0.00565	0.23131	4.32324	0.00131	764.5981
26	217.53477	0.00460	0.23106	4.32784	0.00106	941.4553
27	267.56763	0.00374	0.23086	4.33158	0.00086	1158.9897
28	329.10791	0.00304	0.23070	4.33461	0.00070	1426.5561
29	404.80273	0.00247	0.23057	4.33709	0.00057	1755.6640
30	497.90723	0.00201	0.23046	4.33909	0.00046	2160.4668
31	612.42554	0.00163	0.23038	4.34073	0.00038	2658.3720
32	753.28320	0.00133	0.23031	4.34205	0.00031	3270.7983
33	926.53809	0.00108	0.23025	4.34313	0.00025	4024.0783
34	1139.64136	0.00088	0.23020	4.34401	0.00020	4950.6132
35	1401.75854	0.00071	0.23016	4.34473	0.00016	6090.2539
36	1724.16235	0.00058	0.23013	4.34530	0.00013	7492.0078
37	2120.71899	0.00047	0.23011	4.34578	0.00011	9216.1679
38	2608.48340	0.00038	0.23009	4.34616	0.00009	11336.8828
39	3208.43335	0.00031	0.23007	4.34647	0.00007	13945.3633
40	3946.37158	0.00025	0.23006	4.34672	0.00006	17153.7891

24.00% Compound Interest Factors

Periods	Single Payment		Uniform Series			
	Compound Amount Factor SCA	Present Worth Factor SPW	Capitol Recovery Factor UCR	Present Worth Factor UPW	Sinking Fund Factor USF	Compound Amount Factor UCA
	1	2	3	4	5	6
1	1.24000	0.80645	1.24000	0.80645	1.00000	1.0000
2	1.53760	0.65036	0.68643	1.45681	0.44643	2.2399
3	1.90662	0.52449	0.50472	1.98130	0.26472	3.7776
4	2.36421	0.42297	0.41593	2.40427	0.17593	5.6842
5	2.93162	0.34111	0.36425	2.74538	0.12425	8.0484
6	3.36521	0.27509	0.33107	3.02047	0.09107	10.9800
7	4.50766	0.22184	0.30842	3.24231	0.06842	14.6152
8	5.58950	0.17891	0.29229	3.42122	0.05229	19.1228
9	6.93098	0.14428	0.28047	3.56550	0.04047	24.7123
10	8.59441	0.11635	0.27160	3.68186	0.03160	31.6433
11	10.65707	0.09383	0.26485	3.77569	0.02485	40.2377
12	13.21476	0.07567	0.25965	3.85136	0.01965	50.8948
13	16.38629	0.06103	0.25560	3.91239	0.01560	64.1095
14	20.31900	0.04922	0.25242	3.96160	0.01242	80.4958
15	25.19556	0.03969	0.24992	4.00129	0.00992	100.8148
16	31.24249	0.03201	0.24794	4.03330	0.00794	126.0104
17	38.74068	0.02581	0.24636	4.05911	0.00636	157.2528
18	48.03844	0.02082	0.24510	4.07993	0.00510	195.9935
19	59.56764	0.01679	0.24410	4.09672	0.00410	244.0320
20	73.86388	0.01354	0.24329	4.11026	0.00329	303.5993
21	91.59119	0.01092	0.24265	4.12118	0.00265	377.4631
22	113.57304	0.00880	0.24213	4.12998	0.00213	469.0542
23	140.83055	0.00710	0.24172	4.13708	0.00172	582.6272
24	174.62985	0.00573	0.24138	4.14281	0.00138	723.4575
25	216.54099	0.00462	0.24111	4.14742	0.00111	898.0874
26	268.51074	0.00372	0.24090	4.15115	0.00090	1114.6279
27	332.95312	0.00300	0.24072	4.15415	0.00072	1383.1379
28	412.86182	0.00242	0.24058	4.15657	0.00058	1716.0908
29	511.94873	0.00195	0.24047	4.15853	0.00047	2128.9531
30	634.81641	0.00158	0.24038	4.16010	0.00038	2640.9016
31	787.17212	0.00127	0.24031	4.16137	0.00031	3275.7177
32	976.09326	0.00102	0.24025	4.16240	0.00025	4062.8908
33	1210.35547	0.00083	0.24020	4.16323	0.00020	5038.9804
34	1500.84058	0.00067	0.24016	4.16389	0.00016	6249.3320
35	1861.04199	0.00054	0.24013	4.16443	0.00013	7750.1718
36	2307.69165	0.00043	0.24010	4.16486	0.00010	9611.2148
37	2861.53711	0.00035	0.24008	4.16521	0.00008	11918.9023
38	3548.30518	0.00028	0.24007	4.16549	0.00007	14780.4375
39	4399.89453	0.00023	0.24005	4.16572	0.00005	18328.7266
40	5455.87109	0.00018	0.24004	4.16590	0.00004	22728.6328

25.00% Compound Interest Factors

Periods	Single Payment		Uniform Series			
	Compound Amount Factor SCA 1	Present Worth Factor SPW 2	Capitol Recovery Factor UCR 3	Present Worth Factor UPW 4	Sinking Fund Factor USF 5	Compound Amount Factor UCA 6
1	1.25000	0.80000	1.25000	0.80000	1.00000	1.0000
2	1.56250	0.64000	0.69444	1.44000	0.44444	2.2500
3	1.95313	0.51200	0.51230	1.95200	0.26230	3.8125
4	2.44141	0.40960	0.42344	2.36160	0.17344	5.7656
5	3.05176	0.32768	0.37185	2.68928	0.12185	8.2070
6	3.81470	0.26214	0.33882	2.95142	0.08882	11.2588
7	4.76837	0.20972	0.31634	3.16114	0.06634	15.0734
8	5.96046	0.16777	0.30040	3.32891	0.05040	19.8418
9	7.45058	0.13422	0.28876	3.46313	0.03876	25.8023
10	9.31323	0.10737	0.28007	3.57050	0.03007	33.2529
11	11.64153	0.08590	0.27349	3.65640	0.02349	42.5661
12	14.55192	0.06872	0.26845	3.72512	0.01845	54.2076
13	18.18988	0.05498	0.26454	3.78010	0.01454	68.7595
14	22.73737	0.04398	0.26150	3.82408	0.01150	86.9494
15	28.42171	0.03518	0.25912	3.85926	0.00912	109.6868
16	35.52713	0.02815	0.25724	3.88741	0.00724	138.1085
17	44.40892	0.02252	0.25576	3.90993	0.00576	173.6357
18	55.51114	0.01801	0.25459	3.92794	0.00459	218.0446
19	69.38893	0.01441	0.25366	3.94235	0.00366	273.5556
20	86.73616	0.01153	0.25292	3.95388	0.00292	342.9445
21	108.42021	0.00922	0.25233	3.96311	0.00233	429.6806
22	135.52527	0.00738	0.25186	3.97049	0.00186	538.1010
23	169.40659	0.00590	0.25148	3.97639	0.00148	673.6262
24	211.75822	0.00472	0.25119	3.98111	0.00119	843.0329
25	264.69775	0.00378	0.25095	3.98489	0.00095	1054.7910
26	330.87207	0.00302	0.25076	3.98791	0.00076	1319.4885
27	413.59009	0.00242	0.25061	3.99033	0.00061	1650.3603
28	516.98779	0.00193	0.25048	3.99226	0.00048	2063.9511
29	646.23462	0.00155	0.25039	3.99381	0.00039	2580.9394
30	807.79346	0.00124	0.25031	3.99505	0.00031	3227.1755
31	1009.74194	0.00099	0.25025	3.99604	0.00025	4034.9699
32	1262.17725	0.00079	0.25020	3.99683	0.00020	5044.7070
33	1577.72168	0.00063	0.25016	3.99746	0.00016	6306.8867
34	1972.15210	0.00051	0.25013	3.99797	0.00013	7884.6054
35	2465.19019	0.00041	0.25010	3.99838	0.00010	9856.7578
36	3081.48779	0.00032	0.25008	3.99870	0.00008	12321.9492
37	3851.85986	0.00026	0.25006	3.99896	0.00006	15403.4375
38	4814.82422	0.00021	0.25005	3.99917	0.00005	19255.3008
39	6018.52734	0.00017	0.25004	3.99934	0.00004	24070.1133
40	7523.16016	0.00013	0.25003	3.99947	0.00003	30088.6406

26.00% Compound Interest Factors

Periods	Single Payment		Uniform Series			
	Compound Amount Factor SCA 1	Present Worth Factor SPW 2	Capitol Recovery Factor UCR 3	Present Worth Factor UPW 4	Sinking Fund Factor USF 5	Compound Amount Factor UCA 6
1	1.26000	0.79365	1.26000	0.79365	1.00000	1.0000
2	1.58760	0.62988	0.70248	1.42353	0.44248	2.2599
3	2.00037	0.49991	0.51990	1.92343	0.25990	3.8475
4	2.52047	0.39675	0.43100	2.32019	0.17100	5.8479
5	3.17579	0.31488	0.37950	2.63507	0.11950	8.3684
6	4.00149	0.24991	0.34662	2.88497	0.08662	11.5442
7	5.04187	0.19834	0.32433	3.08331	0.06433	15.5456
8	6.35276	0.15741	0.30857	3.24072	0.04857	20.5875
9	8.00447	0.12493	0.29712	3.36565	0.03712	26.9402
10	10.08563	0.09915	0.28862	3.46480	0.02862	34.9447
11	12.70788	0.07869	0.28221	3.54350	0.02221	45.0303
12	16.01192	0.06245	0.27732	3.60595	0.01732	57.7381
13	20.17500	0.04957	0.27356	3.65551	0.01356	73.7500
14	25.42050	0.03934	0.27065	3.69485	0.01065	93.9250
15	32.02980	0.03122	0.26838	3.72604	0.00838	119.3454
16	40.35753	0.02478	0.26661	3.75085	0.00661	151.3752
17	50.85046	0.01967	0.26522	3.77052	0.00522	191.7326
18	64.07155	0.01561	0.26412	3.78613	0.00412	242.5829
19	80.73012	0.01239	0.26326	3.79851	0.00326	306.6543
20	101.71989	0.00983	0.26258	3.80834	0.00258	387.3840
21	128.16699	0.00780	0.26204	3.81614	0.00204	489.1037
22	161.49031	0.00619	0.26162	3.82234	0.00162	617.2702
23	203.47768	0.00491	0.26128	3.82725	0.00128	778.7602
24	256.38159	0.00390	0.26102	3.83115	0.00102	982.2370
25	323.04077	0.00310	0.26081	3.83425	0.00081	1238.6186
26	407.03101	0.00246	0.26064	3.83671	0.00064	1561.6579
27	512.85889	0.00195	0.26051	3.83865	0.00051	1968.6882
28	646.20190	0.00155	0.26040	3.84020	0.00040	2481.5468
29	814.21387	0.00123	0.26032	3.84143	0.00032	3127.7478
30	1025.90894	0.00097	0.26025	3.84241	0.00025	3941.9606
31	1292.64453	0.00077	0.26020	3.84318	0.00020	4967.8632
32	1628.73120	0.00061	0.26016	3.84379	0.00016	6260.5039
33	2052.20020	0.00049	0.26013	3.84428	0.00013	7889.2304
34	2585.77075	0.00039	0.26010	3.84467	0.00010	9941.4257
35	3258.06934	0.00031	0.26008	3.84497	0.00008	12527.1914
36	4105.16406	0.00024	0.26006	3.84522	0.00006	15785.2500
37	5172.50391	0.00019	0.26005	3.84541	0.00005	19890.4023
38	6517.35156	0.00015	0.26004	3.84556	0.00004	25062.8945
39	8211.85937	0.00012	0.26003	3.84569	0.00003	31580.2344
40	*********	0.00010	0.26003	3.84578	0.00003	39792.0703

27.00% Compound Interest Factors

	Single Payment		Uniform Series			
Periods	Compound Amount Factor SCA 1	Present Worth Factor SPW 2	Capitol Recovery Factor UCR 3	Present Worth Factor UPW 4	Sinking Fund Factor USF 5	Compound Amount Factor UCA 6
1	1.27000	0.78740	1.27000	0.78740	1.00000	1.0000
2	1.61290	0.62000	0.71053	1.40740	0.44053	2.2699
3	2.04838	0.48819	0.52754	1.89559	0.25754	3.8828
4	2.60144	0.38440	0.43860	2.27999	0.16860	5.9312
5	3.30383	0.30268	0.38720	2.58267	0.11720	8.5327
6	4.19586	0.23833	0.35448	2.82100	0.08448	11.8365
7	5.32874	0.18766	0.33237	3.00866	0.06237	16.0323
8	6.76750	0.14776	0.31681	3.15643	0.04681	21.3611
9	8.59472	0.11635	0.30555	3.27278	0.03555	28.1286
10	10.91530	0.09161	0.29723	3.36439	0.02723	36.7233
11	13.86242	0.07214	0.29099	3.43653	0.02099	47.6386
12	17.60526	0.05680	0.28626	3.49333	0.01626	61.5009
13	22.35867	0.04473	0.28264	3.53805	0.01264	79.1062
14	28.39551	0.03522	0.27986	3.57327	0.00986	101.4648
15	36.06229	0.02773	0.27770	3.60100	0.00770	129.8603
16	45.79907	0.02183	0.27603	3.62284	0.00603	165.9225
17	58.16481	0.01719	0.27472	3.64003	0.00472	211.7215
18	73.86928	0.01354	0.27371	3.65357	0.00371	269.8862
19	93.81395	0.01066	0.27291	3.66423	0.00291	343.7553
20	119.14368	0.00839	0.27229	3.67262	0.00229	437.5690
21	151.31241	0.00661	0.27180	3.67923	0.00180	556.7126
22	192.16669	0.00520	0.27141	3.68443	0.00141	708.0246
23	244.05161	0.00410	0.27111	3.68853	0.00111	900.1911
24	309.94531	0.00323	0.27087	3.69175	0.00087	1144.2419
25	393.63037	0.00254	0.27069	3.69429	0.00069	1454.1867
26	499.91040	0.00200	0.27054	3.69629	0.00054	1847.8166
27	634.88599	0.00158	0.27043	3.69787	0.00043	2347.7270
28	806.30493	0.00124	0.27034	3.69911	0.00034	2982.6118
29	1024.00708	0.00098	0.27026	3.70009	0.00026	3788.9157
30	1300.48853	0.00077	0.27021	3.70086	0.00021	4812.9179
31	1651.61963	0.00061	0.27016	3.70146	0.00016	6113.4062
32	2097.55615	0.00048	0.27013	3.70194	0.00013	7765.0234
33	2663.89551	0.00038	0.27010	3.70231	0.00010	9862.5742
34	3383.14600	0.00030	0.27008	3.70261	0.00008	12526.4648
35	4296.59375	0.00023	0.27006	3.70284	0.00006	15909.6094
36	5456.67187	0.00018	0.27005	3.70302	0.00005	20206.1953
37	6929.96875	0.00014	0.27004	3.70317	0.00004	25662.8516
38	8801.05859	0.00011	0.27003	3.70328	0.00003	32592.8164
39	*********	0.00009	0.27002	3.70337	0.00002	41393.8555
40	*********	0.00007	0.27002	3.70344	0.00002	52571.1836

28.00% Compound Interest Factors

Periods	Single Payment		Uniform Series			
	Compound Amount Factor SCA	Present Worth Factor SPW	Capitol Recovery Factor UCR	Present Worth Factor UPW	Sinking Fund Factor USF	Compound Amount Factor UCA
	1	2	3	4	5	6
1	1.28000	0.78125	1.28000	0.78125	1.00000	1.0000
2	1.63840	0.61035	0.71860	1.39160	0.43860	2.2800
3	2.09715	0.47684	0.53521	1.86844	0.25521	3.9183
4	2.68435	0.37253	0.44624	2.24097	0.16624	6.0155
5	3.43597	0.29104	0.39494	2.53200	0.11494	8.6998
6	4.39804	0.22737	0.36240	2.75938	0.08240	12.1358
7	5.62949	0.17764	0.34048	2.93702	0.06048	16.5338
8	7.20575	0.13878	0.32512	3.07579	0.04512	22.1633
9	9.22335	0.10842	0.31405	3.18421	0.03405	29.3691
10	11.80589	0.08470	0.30591	3.26892	0.02591	38.5924
11	15.11154	0.06617	0.29984	3.33509	0.01984	50.3983
12	19.34276	0.05170	0.29526	3.38679	0.01526	65.5098
13	24.75873	0.04039	0.29179	3.42718	0.01179	84.8526
14	31.69116	0.03155	0.28912	3.45873	0.00912	109.6113
15	40.56468	0.02465	0.28708	3.48339	0.00708	141.3024
16	51.92279	0.01926	0.28550	3.50265	0.00550	181.8672
17	66.46115	0.01505	0.28428	3.51769	0.00428	233.7898
18	85.07027	0.01175	0.28333	3.52945	0.00333	300.2509
19	108.88992	0.00918	0.28260	3.53863	0.00260	385.3210
20	139.37907	0.00717	0.28202	3.54581	0.00202	494.2109
21	178.40517	0.00561	0.28158	3.55141	0.00158	633.5898
22	228.35858	0.00438	0.28123	3.55579	0.00123	811.9948
23	292.29883	0.00342	0.28096	3.55921	0.00096	1040.3530
24	374.14233	0.00267	0.28075	3.56188	0.00075	1332.6516
25	478.90234	0.00209	0.28059	3.56397	0.00059	1706.7944
26	612.99487	0.00163	0.28046	3.56560	0.00046	2185.6972
27	784.63306	0.00127	0.28036	3.56688	0.00036	2798.6897
28	1004.33032	0.00100	0.28028	3.56787	0.00028	3583.3259
29	1285.54248	0.00078	0.28022	3.56865	0.00022	4587.6523
30	1645.49414	0.00061	0.28017	3.56926	0.00017	5873.1914
31	2106.23218	0.00047	0.28013	3.56973	0.00013	7518.6875
32	2695.97656	0.00037	0.28010	3.57010	0.00010	9624.9140
33	3450.84937	0.00029	0.28008	3.57039	0.00008	12320.8906
34	4417.08594	0.00023	0.28006	3.57062	0.00006	15771.7383
35	5653.86719	0.00018	0.28005	3.57080	0.00005	20188.8125
36	7236.94922	0.00014	0.28004	3.57094	0.00004	25842.6836
37	9263.29297	0.00011	0.28003	3.57104	0.00003	33079.6211
38	*********	0.00008	0.28002	3.57113	0.00002	42342.9219
39	*********	0.00007	0.28002	3.57119	0.00002	54199.9570
40	*********	0.00005	0.28001	3.57125	0.00001	69376.8750

29.00% Compound Interest Factors

Periods	Single Payment		Uniform Series			
	Compound Amount Factor SCA	Present Worth Factor SPW	Capitol Recovery Factor UCR	Present Worth Factor UPW	Sinking Fund Factor USF	Compound Amount Factor UCA
	1	2	3	4	5	6
1	1.29000	0.77519	1.29000	0.77519	1.00000	1.0000
2	1.66410	0.60093	0.72668	1.37612	0.43668	2.2900
3	2.14669	0.46583	0.54290	1.84195	0.25290	3.9541
4	2.76923	0.36111	0.45391	2.20306	0.16391	6.1007
5	3.57230	0.27993	0.40274	2.48300	0.11274	8.8700
6	4.60827	0.21700	0.37037	2.70000	0.08037	12.4423
7	5.94467	0.16822	0.34865	2.86821	0.05865	17.0505
8	7.66863	0.13040	0.33349	2.99862	0.04349	22.9952
9	9.89253	0.10109	0.32261	3.09970	0.03261	30.6638
10	12.76136	0.07836	0.31466	3.17806	0.02466	40.5564
11	16.46214	0.06075	0.30876	3.23881	0.01876	53.3177
12	21.23618	0.04709	0.30433	3.28590	0.01433	69.7799
13	27.39467	0.03650	0.30099	3.32240	0.01099	91.0161
14	35.33911	0.02830	0.29845	3.35070	0.00845	118.4107
15	45.58746	0.02194	0.29650	3.37263	0.00650	153.7498
16	58.80782	0.01700	0.29502	3.38964	0.00502	199.3373
17	75.86209	0.01318	0.29387	3.40282	0.00387	258.1450
18	97.86209	0.01022	0.29299	3.41304	0.00299	334.0070
19	126.24211	0.00792	0.29232	3.42096	0.00232	431.8691
20	162.85231	0.00614	0.29179	3.42710	0.00179	558.1113
21	210.07948	0.00476	0.29139	3.43186	0.00139	720.9636
22	271.00244	0.00369	0.29107	3.43555	0.00107	931.0427
23	349.59302	0.00286	0.29083	3.43841	0.00083	1202.0449
24	450.97510	0.00222	0.29064	3.44063	0.00064	1551.6381
25	581.75806	0.00172	0.29050	3.44235	0.00050	2002.6145
26	750.46777	0.00133	0.29039	3.44368	0.00039	2584.3715
27	968.10352	0.00103	0.29030	3.44471	0.00030	3334.8400
28	1248.85352	0.00080	0.29023	3.44551	0.00023	4302.9414
29	1611.02100	0.00062	0.29018	3.44614	0.00018	5551.7929
30	2078.21704	0.00048	0.29014	3.44662	0.00014	7162.8164
31	2680.89990	0.00037	0.29011	3.44699	0.00011	9241.0312
32	3458.36084	0.00029	0.29008	3.44728	0.00008	11921.9297
33	4461.28516	0.00022	0.29006	3.44750	0.00007	15380.2930
34	5755.05469	0.00017	0.29005	3.44768	0.00005	19841.5703
35	7424.02344	0.00013	0.29004	3.44781	0.00004	25596.6367
36	9576.98828	0.00010	0.29003	3.44792	0.00003	33020.6484
37	*********	0.00008	0.29002	3.44800	0.00002	42597.6562
38	*********	0.00006	0.29002	3.44806	0.00002	54951.9727
39	*********	0.00005	0.29001	3.44811	0.00001	70889.0000
40	*********	0.00004	0.29001	3.44815	0.00001	91447.8125

30.00% Compound Interest Factors

Periods	Single Payment		Uniform Series			
	Compound Amount Factor SCA	Present Worth Factor SPW	Capitol Recovery Factor UCR	Present Worth Factor UPW	Sinking Fund Factor USF	Compound Amount Factor UCA
	1	2	3	4	5	6
1	1.30000	0.76923	1.30000	0.76923	1.00000	1.0000
2	1.69000	0.59172	0.73478	1.36094	0.43478	2.2999
3	2.19700	0.45517	0.55063	1.81611	0.25063	3.9899
4	2.85609	0.35013	0.46163	2.16624	0.16163	6.1869
5	3.71292	0.26933	0.41058	2.43557	0.11058	9.0430
6	4.82679	0.20718	0.37839	2.64274	0.07839	12.7559
7	6.27483	0.15937	0.35687	2.80211	0.05687	17.5827
8	8.15727	0.12259	0.34192	2.92470	0.04192	23.8575
9	10.60444	0.09430	0.33124	3.01900	0.03124	32.0148
10	13.78577	0.07254	0.32346	3.09154	0.02346	42.6192
11	17.92148	0.05580	0.31773	3.14734	0.01773	56.4049
12	23.29791	0.04292	0.31345	3.19026	0.01345	74.3263
13	30.28728	0.03302	0.31024	3.22328	0.01024	97.6242
14	39.37343	0.02540	0.30782	3.24867	0.00782	127.9114
15	51.18544	0.01954	0.30598	3.26821	0.00598	167.2848
16	66.54103	0.01503	0.30458	3.28324	0.00458	218.4702
17	86.50330	0.01156	0.30351	3.29480	0.00351	285.0107
18	112.45421	0.00889	0.30269	3.30369	0.00269	371.5139
19	146.19038	0.00684	0.30207	3.31053	0.00207	483.9677
20	190.04739	0.00526	0.30159	3.31579	0.00159	630.1579
21	247.06148	0.00405	0.30122	3.31984	0.00122	820.2048
22	321.17969	0.00311	0.30094	3.32296	0.00094	1067.2656
23	417.53320	0.00240	0.30072	3.32535	0.00072	1388.4440
24	542.79297	0.00184	0.30055	3.32719	0.00055	1805.9768
25	705.63062	0.00142	0.30043	3.32861	0.00043	2348.7695
26	917.31909	0.00109	0.30033	3.32970	0.00033	3054.3986
27	1192.51416	0.00084	0.30025	3.33054	0.00025	3971.7170
28	1550.26758	0.00065	0.30019	3.33118	0.00019	5164.2226
29	2015.34692	0.00050	0.30015	3.33168	0.00015	6714.4882
30	2619.94946	0.00038	0.30011	3.33206	0.00011	8729.8281
31	3405.93213	0.00029	0.30009	3.33235	0.00009	11349.7734
32	4427.70703	0.00023	0.30007	3.33258	0.00007	14755.6875
33	5756.01562	0.00017	0.30005	3.33275	0.00005	19183.3867
34	7482.81641	0.00013	0.30004	3.33289	0.00004	24939.3906
35	9727.66016	0.00010	0.30003	3.33299	0.00003	32422.2031
36	*********	0.00008	0.30002	3.33307	0.00002	42149.8516
37	*********	0.00006	0.30002	3.33313	0.00002	54795.7617
38	*********	0.00005	0.30001	3.33318	0.00001	71235.3750
39	*********	0.00004	0.30001	3.33321	0.00001	92607.0000
40	*********	0.00003	0.30001	3.33324	0.00001	*********

Discount-Escalation Factors

Present Worth of a Series of Escalating Payments Compounded Annually

DISCOUNT-ESCALATION FACTORS FOR N = 2 YEARS

Discount Rate	Annual Escalation Rate					
	.05	.06	.07	.08	.09	.10
0.06	1.971929	2.000000	2.028252	2.056846	2.085643	2.114577
0.07	1.944315	1.972155	2.000000	2.028052	2.056394	2.084881
0.08	1.917439	1.944882	1.972321	2.000000	2.027691	2.055803
0.09	1.891293	1.918221	1.945303	1.972645	2.000000	2.027536
0.10	1.865720	1.892277	1.918934	1.945849	1.972856	2.000000
0.11	1.840763	1.866907	1.893195	1.919710	1.946316	1.973215
0.12	1.816424	1.842180	1.868067	1.894139	1.920402	1.946800
0.13	1.792635	1.818007	1.843534	1.869234	1.895102	1.921083
0.14	1.769395	1.794416	1.819564	1.844899	1.870348	1.895968
0.15	1.746706	1.771347	1.796147	1.821124	1.846208	1.871487
0.16	1.724516	1.748822	1.773264	1.797866	1.822617	1.847519
0.17	1.702830	1.726792	1.750900	1.775162	1.799563	1.824102
0.18	1.681640	1.705267	1.729033	1.752961	1.777022	1.801208
0.19	1.660906	1.684204	1.707652	1.731236	1.754961	1.778845
0.20	1.640627	1.663617	1.686739	1.710006	1.733406	1.756952
0.21	1.620801	1.643469	1.666285	1.689240	1.712322	1.735543
0.22	1.601390	1.623763	1.646269	1.668919	1.691691	1.714594
0.23	1.582394	1.604468	1.626680	1.649021	1.671503	1.694107
0.24	1.563807	1.585591	1.607508	1.629558	1.651731	1.674044
0.25	1.545603	1.567110	1.588738	1.610503	1.632385	1.654402
0.26	1.527781	1.549008	1.570361	1.591838	1.613447	1.635174
0.27	1.510323	1.531285	1.552362	1.573566	1.594897	1.616351
0.28	1.493229	1.513918	1.534732	1.555669	1.576728	1.597905
0.29	1.476475	1.496910	1.517459	1.538135	1.558920	1.579834
0.30	1.460059	1.480239	1.500536	1.520948	1.541482	1.562131

Present Worth of a Series of Escalating Payments Compounded Annually

DISCOUNT-ESCALATION FACTORS FOR N = 3 YEARS

Discount Rate	Annual Escalation Rate					
	.05	.06	.07	.08	.09	.10
0.06	2.943965	3.000000	3.056715	3.114442	3.172973	3.232069
0.07	2.889300	2.944430	3.000000	3.056344	3.113490	3.171347
0.08	2.836400	2.890394	2.944802	3.000000	3.055727	3.112371
0.09	2.785209	2.837917	2.891270	2.945412	3.000000	3.055257
0.10	2.735469	2.787126	2.839331	2.892320	2.945848	3.000000
0.11	2.687209	2.737774	2.788937	2.840829	2.893250	2.946508
0.12	2.640407	2.689932	2.740030	2.790780	2.842195	2.894207
0.13	2.594932	2.643444	2.692555	2.742288	2.792642	2.843545
0.14	2.550760	2.598326	2.646436	2.695181	2.744457	2.794357
0.15	2.507869	2.554461	2.601635	2.649418	2.697713	2.746655
0.16	2.466160	2.511861	2.558098	2.604913	2.652291	2.700241
0.17	2.425619	2.470429	2.515784	2.561695	2.608149	2.655145
0.18	2.386211	2.430160	2.474635	2.519669	2.565226	2.611296
0.19	2.347861	2.390973	2.434613	2.478770	2.523451	2.568688
0.20	2.310549	2.352864	2.395677	2.439009	2.482845	2,527210
0.21	2.274255	2.315767	2.357792	2.400320	2.443336	2.486861
0.22	2.238904	2.279666	2.320910	2.362657	2.404877	2.447586
0.23	2.204483	2.244501	2.285001	2.325972	2.367435	2.409369
0.24	2.170970	2.210265	2.250029	2.290263	2.330956	2.372138
0.25	2.138309	2.176912	2.215961	2.255478	2.295441	2.335874
0.26	2.106484	2.144406	2.182770	2.221577	2.260841	2.300550
0.27	2.075463	2.112729	2.150417	2.188547	2.227120	2.266135
0.28	2.045229	2.081840	2.118879	2.156348	2.194248	2.232576
0.29	2.015736	2.051726	2.088126	2.124953	2.162188	2.199859
0.30	1.986971	2.022350	2.058136	2.094328	2.130937	2.167958

Present Worth of a Series of Escalating Payments Compounded Annually

DISCOUNT-ESCALATION FACTORS FOR N = 4 YEARS

Discount Rate	Annual Escalation Rate					
	.05	.06	.07	.08	.09	.10
0.06	3.906825	4.000000	4.094974	4.192019	4.291037	4.391757
0.07	3.816621	3.907615	4.000000	4.094213	4.190389	4.288304
0.08	3.729833	3.818395	3.908284	4.000000	4.093224	4.188503
0.09	3.646319	3.732302	3.819876	3.909256	4.000000	4.092418
0.10	3.565682	3.649434	3.734625	3.821580	3.909998	4.000000
0.11	3.487902	3.569413	3.652401	3.737054	3.823118	3.911030
0.12	3.412889	3.492270	3.573066	3.655398	3.739299	3.825692
0.13	3.340429	3.417750	3.496494	3.576711	3.658411	3.741515
0.14	3.270438	3.345820	3.422535	3.500710	3.580228	3.661222
0.15	3.202844	3.276288	3.351088	3.427294	3.504792	3.583772
0.16	3.137474	3.209120	3.282040	3.356301	3.431898	3.508858
0.17	3.074275	3.144153	3.215293	3.287727	3.361446	3.436466
0.18	3.013159	3.081335	3.150731	3.221401	3.293310	3.366463
0.19	2.953998	3.020531	3.088268	3.157205	3.227363	3.298796
0.20	2.896730	2.961699	3.027813	3.095111	3.163585	3.233280
0.21	2.841300	2.904723	2.969291	3.035003	3.101852	3.169877
0.22	2.787584	2.849549	2.912604	2.976786	3.042068	3.108479
0.23	2.735535	2.796074	2.857685	2.920368	2.984155	3.049035
0.24	2.685099	2.744260	2.804461	2.865716	2.928017	2.991415
0.25	2.636180	2.694024	2.752864	2.812737	2.873624	2.935570
0.26	2.588738	2.645296	2.702831	2.761353	2.820889	2.881433
0.27	2.542706	2.598029	2.654290	2.711523	2.769736	2.828939
0.28	2.498040	2.552149	2.607189	2.663172	2.720105	2.777997
0.29	2.454668	2.507622	2.561470	2.616243	2.671925	2.728561
0.30	2.412554	2.464379	2.517083	2.570675	2.625171	2.680580

Present Worth of a Series of Escalating Payments Compounded Annually

DISCOUNT-ESCALATION FACTORS FOR N = 5 YEARS

Discount Rate	Annual Escalation Rate					
	.05	.06	.07	.08	.09	.10
0.06	4.860604	5.000000	5.143028	5.289938	5.440772	5.595191
0.07	4.726605	4.861801	5.000000	5.141868	5.287403	5.436560
0.08	4.598448	4.729209	4.862844	5.000000	5.140388	5.284516
0.09	4.475832	4.602073	4.731446	4.864255	5.000000	5.139125
0.10	4.358160	4.480386	4.605503	4.733947	4.865384	5.000000
0.11	4.245313	4.363589	4.484749	4.609056	4.736236	4.866862
0.12	4.137092	4.251623	4.368912	4.489139	4.612371	4.738562
0.13	4.033150	4.144090	4.257748	4.374213	4.493533	4.615647
0.14	3.933299	4.040860	4.150979	4.263844	4.379341	4.497670
0.15	3.837385	3.941625	4.048406	4.157821	4.269762	4.384494
0.16	3.745130	3.846270	3.949814	4.055870	4.164460	4.275647
0.17	3.656403	3.754535	3.855016	3.957911	4.063236	4.171042
0.18	3.571037	3.666289	3.763802	3.863665	3.965864	4.070432
0.19	3.488824	3.581314	3.676008	3.772927	3.872122	3.973684
0.20	3.409638	3.499504	3.591468	3.685602	3.781923	3.880510
0.21	3.333365	3.420668	3.510037	3.601498	3.695062	3.790801
0.22	3.259809	3.344693	3.431549	3.520441	3.611360	3.704368
0.23	3.188872	3.271413	3.355874	3.442276	3.530679	3.621094
0.24	3.120449	3.200740	3.282884	3.366917	3.452853	3.540773
0.25	3.054391	3.132535	3.212452	3.294209	3.377799	3.463301
0.26	2.990616	3.066680	3.144470	3.224019	3.305375	3.388553
0.27	2.929008	3.003083	3.078814	3.156258	3.235445	3.316408
0.28	2.869488	2.941624	3.015387	3.090804	3.167905	3.246718
0.29	2.811938	2.882234	2.954088	3.027556	3.102633	3.179393
0.30	2.756293	2.824803	2.894833	2.966409	3.039568	3.114338

Present Worth of a Series of Escalating Payments Compounded Annually

DISCOUNT-ESCALATION FACTORS FOR N = 6 YEARS

Discount Rate	Annual Escalation Rate					
	.05	.06	.07	.08	.09	.10
0.06	5.805389	6.00000	6.200878	6.408558	6.623044	6.844022
0.07	5.619586	5.807066	6.000000	6.199306	6.404896	6.617020
0.08	5.442934	5.623158	5.808558	6.000000	6.197116	6.400881
0.09	5.274902	5.447904	5.626287	5.810497	6.000000	6.195483
0.10	5.114615	5.281123	5.452630	5.629728	5.812084	6.000000
0.11	4.961782	5.121992	5.287102	5.457490	5.632899	5.814086
0.12	4.816031	4.970297	5.129230	5.293100	5.462058	5.636113
0.13	4.676826	4.825435	4.978580	5.136429	5.299093	5.466572
0.14	4.543830	4.687125	4.834694	4.986814	5.143407	5.304771
0.15	4.416749	4.554892	4.697214	4.843881	4.994820	5.150400
0.16	4.295162	4.428493	4.565778	4.707193	4.852815	5.002776
0.17	4.178824	4.307530	4.440061	4.576542	4.717040	4.861670
0.18	4.067454	4.191757	4.319721	4.451500	4.587121	4.726675
0.19	3.960729	4.080834	4.204480	4.331735	4.462698	4.597531
0.20	3.858433	3.974565	4.094060	4.217046	4.343580	4.473804
0.21	3.760364	3.872653	3.988219	4.107130	4.229440	4.355278
0.22	3.666232	3.774900	3.886690	4.001710	4.119990	4.241644
0.23	3.575867	3.681055	3.789259	3.900538	4.014997	4.132690
0.24	3.489093	3.590956	3.695715	3.803447	3.914200	4.028108
0.25	3.405689	3.504393	3.605860	3.710199	3.817440	3.927704
0.26	3.325515	3.421177	3.519513	3.620589	3.724493	3.831278
0.27	3.248391	3.341159	3.436482	3.534458	3.635149	3.738626
0.28	3.174190	3.264157	3.356613	3.451619	3.549234	3.649526
0.29	3.102740	3.190054	3.279748	3.371911	3.466563	3.563824
0.30	3.033929	3.118687	3.205750	3.295173	3.387024	3.481363

Present Worth of a Series of Escalating Payments Compounded Annually

DISCOUNT-ESCALATION FACTORS FOR N = 7 YEARS

Discount Rate	Annual Escalation Rate					
	.05	.06	.07	.08	.09	.10
0.06	6.741256	7.000000	7.268728	7.548292	7.838788	8.140007
0.07	6.495872	6.743498	7.000000	7.266528	7.543335	7.830595
0.08	6.263960	6.500552	6.745522	7.000000	7.263721	7.537911
0.09	6.044648	6.270453	6.504713	6.748062	7.000000	7.261493
0.10	5.836686	6.052741	6.276654	6.509221	6.750175	7.000000
0.11	5.639524	5.846233	6.060543	6.282994	6.513409	6.752773
0.12	5.452536	5.650470	5.855604	6.068350	6.288983	6.517633
0.13	5.274933	5.464573	5.661139	5.864919	6.076138	6.294906
0.14	5.106160	5.288038	5.476426	5.671732	5.873959	6.083550
0.15	4.945733	5.120162	5.300889	5.488181	5.682049	5.883005
0.16	4.793036	4.960524	5.133951	5.313597	5.499630	5.692293
0.17	4.647665	4.808535	4.975102	5.147588	5.326139	5.510977
0.18	4.509180	4.663787	4.823818	4.989519	5.160993	5.338427
0.19	4.377116	4.525785	4.679662	4.838886	5.003647	5.174196
0.20	4.251128	4.394200	4.542205	4.695345	4.853752	5.017657
0.21	4.130899	4.268607	4.411074	4.558438	4.710822	4.868438
0.22	4.016022	4.148687	4.285871	4.427752	4.574421	4.726073
0.23	3.906228	4.034080	4.166267	4.302914	4.444188	4.590216
0.24	3.801250	3.924528	4.051949	4.183650	4.319738	4.460421
0.25	3.700779	3.819727	3.942616	4.069615	4.200807	4.336379
0.26	3.604596	3.719405	3.838002	3.960507	4.087064	4.217782
0.27	3.512448	3.623333	3.737826	3.856077	3.978201	4.104325
0.28	3.424141	3.531256	3.641858	3.756057	3.873960	3.995687
0.29	3.339438	3.442991	3.549870	3.660208	3.774071	3.891632
0.30	3.258174	3.358315	3.461658	3.568299	3.678351	3.791924

Present Worth of a Series of Escalating Payments Compounded Annually

DISCOUNT-ESCALATION FACTORS FOR N = 8 YEARS

Discount Rate	Annual Escalation Rate					
	.05	.06	.07	.08	.09	.10
0.06	7.668291	8.000000	8.346681	8.709553	9.088906	9.484878
0.07	7.355777	7.671180	8.000000	8.343844	8.702980	9.078160
0.08	7.062181	7.361696	7.673807	8.000000	8.340200	8.695976
0.09	6.786149	7.070368	7.367021	7.677018	8.000000	8.337258
0.10	6.525935	6.796300	7.078206	7.372724	7.679736	8.000000
0.11	6.280629	6.537850	6.806110	7.086189	7.378050	7.683008
0.12	6.049259	6.294206	6.549550	6.815913	7.093760	7.383415
0.13	5.830696	6.064117	6.307456	6.561174	6.825677	7.101249
0.14	5.624096	5.846781	6.078755	6.320603	6.572470	6.835007
0.15	5.428719	5.641195	5.862568	6.093264	6.333422	6.583760
0.16	5.243697	5.446691	5.658042	5.878180	6.107413	6.346147
0.17	5.068419	5.262436	5.464415	5.674706	5.893592	6.121437
0.18	4.902242	5.087813	5.280924	5.481942	5.691095	5.908703
0.19	4.744515	4.922129	5.106925	5.299160	5.499137	5.707247
0.20	4.594736	4.764879	4.941800	5.125813	5.317157	5.516189
0.21	4.452439	4.615476	4.785004	4.961258	5.144464	5.334948
0.22	4.317071	4.473454	4.635971	4.804903	4.980431	5.162854
0.23	4.188243	4.338312	4.494235	4.656219	4.824529	4.999385
0.24	4.065578	4.209678	4.359344	4.514795	4.676220	4.843923
0.25	3.948654	4.087131	4.230880	4.380152	4.535101	4.696012
0.26	3.837165	3.970296	4.108465	4.251866	4.400716	4.555208
0.27	3.730763	3.858848	3.991714	4.129580	4.272632	4.421072
0.28	3.629179	3.752446	3.880304	4.012926	4.150483	4.293170
0.29	3.532101	3.650831	3.773923	3.901572	4.033902	4.171158
0.30	3.439294	3.553704	3.672291	3.795205	3.922619	4.054706

Present Worth of a Series of Escalating Payments Compounded Annually

DISCOUNT-ESCALATION FACTORS FOR N = 9 YEARS

Discount Rate	Annual Escalation Rate					
	.05	.06	.07	.08	.09	.10
0.06	8.586583	9.000000	9.434837	9.892699	10.374436	10.880496
0.07	8.199610	8.590193	9.000000	9.431151	9.884352	10.360728
0.08	7.838227	8.206893	8.593494	9.000000	9.426661	9.875495
0.09	7.500436	7.848263	8.213506	8.597458	9.000000	9.422884
0.10	7.183854	7.512821	7.857894	8.220525	8.600851	9.000000
0.11	6.887081	7.198313	7.524812	7.867674	8.227111	8.604856
0.12	6.608686	6.903454	7.212518	7.536776	7.876979	8.233733
0.13	6.347113	6.626524	6.919455	7.226623	7.548683	7.886183
0.14	6.101142	6.366315	6.644098	6.935323	7.240344	7.560095
0.15	5.869705	6.121450	6.385175	6.661515	6.950809	7.254045
0.16	5.651623	5.890945	6.141470	6.403826	6.678520	6.966180
0.17	5.446019	5.673663	5.911905	6.161277	6.422244	6.695374
0.18	5.252000	5.468719	5.695417	5.932636	6.180766	6.440317
0.19	5.068692	5.175174	5.491102	5.716887	5.952990	6.199984
0.20	4.895392	5.092312	5.298106	5.513235	5.738084	5.973175
0.21	4.731461	4.919344	5.115668	5.320802	5.535100	5.759048
0.22	4.576170	4.755627	4.943027	5.138775	5.343176	5.556672
0.23	4.428988	4.600495	4.779540	4.966439	5.161578	5.365309
0.24	4.289402	4.453435	4.624597	4.803212	4.989578	5.184127
0.25	4.156869	4.313889	4.477633	4.648455	4.826607	5.012490
0.26	4.030972	4.181362	4.338144	4.501602	4.672050	4.849785
0.27	3.911261	4.055419	4.205619	4.362165	4.525331	4.695420
0.28	3.797374	3.935620	4.079631	4.229659	4.385961	4.548820
0.29	3.688917	3.821614	3.959766	4.103645	4.253450	4.409513
0.30	3.585584	3.713022	3.845657	3.983711	4.127428	4.277060

Present Worth of a Series of Escalating Payments Compounded Annually

DISCOUNT-ESCALATION FACTORS FOR N = 10 YEARS

Discount Rate	Annual Escalation Rate					
	.05	.06	.07	.08	.09	.10
0.06	9.496213	10.000000	10.533197	11.098195	11.696315	12.328775
0.07	9.027671	9.500612	10.000000	10.528654	11.087762	11.679244
0.08	8.592717	9.036440	9.504674	10.000000	10.523205	11.076834
0.09	8.188514	8.604750	9.044459	9.509451	10.000000	10.518476
0.10	7.811868	8.203285	8.616322	9.052914	9.513591	10.000000
0.11	7.460752	7.829025	8.217613	8.628040	9.060875	9.518405
0.12	7.133150	7.480064	7.845888	8.231894	8.639216	9.068870
0.13	6.826969	7.154090	7.498959	7.862627	8.246099	8.650280
0.14	6.540526	6.849390	7.174727	7.517690	7.878925	8.259741
0.15	6.272345	6.564121	6.871425	7.195177	7.535986	7.895187
0.16	6.020866	6.296903	6.587391	6.893220	7.215162	7.554141
0.17	5.784891	6.046228	6.321150	6.610420	6.914749	7.234974
0.18	5.563226	5.810889	6.071272	6.345136	6.633089	6.935890
0.19	5.354730	5.589650	5.836540	6.096001	6.368703	6.655455
0.20	5.158467	5.381544	5.615812	5.861916	6.120425	6.392080
0.21	4.973587	5.185543	5.408075	5.641715	5.886995	6.144593
0.22	4.799165	5.000793	5.212331	5.434334	5.667268	5.911755
0.23	4.634502	4.826443	5.027732	5.238828	5.460265	5.692557
0.24	4.478931	4.661808	4.853483	5.054414	5.265031	5.485921
0.25	4.331770	4.506179	4.688854	4.880268	5.080799	5.290990
0.26	4.192477	4.358926	4.533189	4.715661	4.906776	5.106959
0.27	4.060490	4.219487	4.385839	4.559955	4.742216	4.933045
0.28	3.935347	4.087311	4.246255	4.412528	4.586484	4.768518
0.29	3.816560	3.961947	4.113915	4.272822	4.438959	4.612762
0.30	3.703740	3.842927	3.988352	4.140317	4.299152	4.465205

Present Worth of a Series of Escalating Payments Compounded Annually

DISCOUNT-ESCALATION FACTORS FOR N = 11 YEARS

Discount Rate	Annual Escalation Rate					
	.05	.06	.07	.08	.09	.10
0.06	10.397262	11.000000	11.641966	12.326401	13.055615	13.831711
0.07	9.840253	10.402528	11.000000	11.636353	12.313729	13.034720
0.08	9.326249	9.850624	10.407411	11.000000	11.629832	12.300413
0.09	8.851339	9.340416	9.860168	10.413080	11.000000	11.624137
0.10	8.411335	8.868644	9.354063	9.870168	10.418029	11.000000
0.11	8.003412	8.431327	8.885449	9.367854	9.879616	10.423723
0.12	7.624833	8.025785	8.450982	8.902187	9.381040	9.889090
0.13	7.272854	7.648976	8.047693	8.470490	8.918824	9.394093
0.14	6.945222	7.298565	7.672775	8.069405	8.489499	8.934839
0.15	6.639973	6.972148	7.323850	7.696357	8,090631	8.508455
0.16	6.355095	6.667864	6.998714	7.348864	7.719420	8.111691
0.17	6.089006	6.383765	6.695415	7.025015	7.373577	7.742290
0.18	5.840163	6.118261	6.412089	6.722677	7.050913	7.397863
0.19	5.607117	5.869773	6.147143	6.440070	6.749482	7.076478
0.20	5.388658	5.637032	5.899100	6.175727	6.467718	6.776076
0.21	5.183697	5.418742	5.666649	5.928152	6.203992	6.495090
0.22	4.991087	5.213807	5.448523	5.695977	5.956825	6.231911
0.23	4,809940	5.021162	5.243639	5.477998	5.724955	5.985219
0.24	4.639420	4.839933	5.050991	5.273202	5.507162	5.753641
0.25	4.478686	4.669242	4.869659	5.080555	5.302455	5.536070
0.26	4.327065	4.508305	4.698822	4.899141	5.109832	5.331470
0.27	4.183868	4.356425	4.537677	4.728153	4.928361	5.138862
0.28	4.048527	4.212929	4.385543	4.566824	4.757242	4.957321
0.29	3.920454	4.077260	4.241775	4.414459	4.595707	4.786076
0.30	3.799174	3.948850	4.105800	4.270420	4.443136	4.624405

Life Cycle Costing: A Practical Guide for Energy Managers

Present Worth of a Series of Escalating Payments Compounded Annually

DISCOUNT-ESCALATION FACTORS FOR N = 12 YEARS

Discount Rate	Annual Escalation Rate					
	.05	.06	.07	.08	.09	.10
0.06	11.289811	12.000000	12.761143	13.577781	14.453413	15.391372
0.07	10.637645	11.296009	12.000000	12.754456	13.562566	14.428244
0.08	10.039402	10.649729	11.301785	12.000000	12.746649	13.546704
0.09	9.489841	10.055836	10.660911	11.308418	12.000000	12.739870
0.10	8.983555	9.509807	10.071686	10.672565	11.314247	12.000000
0.11	8.516740	9.006497	9.529219	10.087675	10.683604	11.320879
0.12	8.085787	8.542271	9.029062	9.548542	10.102990	10.694662
0.13	7.687171	8.113205	8.567291	9.051456	9.567736	10.118156
0.14	7.317968	7.716219	8.140241	8.592084	9.073292	9.586248
0.15	6.975632	7.348243	7.744802	8.167028	8.616337	9.095058
0.16	6.657630	7.006845	7.378125	7.773084	8.193251	8.640402
0.17	6.361931	6.689569	7.037693	7.407717	7.801034	8.219253
0.18	6.086592	6.394376	6.721135	7.068224	7.436869	7.828518
0.19	5.829810	6.119293	6.426425	6.752336	7.098264	7.465659
0.20	5.590074	5.862714	6.151698	6.458158	6.783175	7.128072
0.21	5.366023	5.623032	5.895307	6.183814	6.489551	6.813722
0.22	5.156266	5.398885	5.655675	5.927595	6.215528	6.520576
0.23	4.959705	5.188968	5.431460	5.688002	5.959517	6.246948
0.24	4.775317	4.992202	5.221420	5.463759	5.720002	5.991135
0.25	4.602096	4.807519	5.024428	5.253603	5.495738	5.751740
0.26	4.439221	4.633973	4.839478	5.056410	5.285492	5.527475
0.27	4.285874	4.470720	4.665605	4.871188	5.088122	5.317128
0.28	4.141370	4.316957	4.501978	4.697011	4.902654	5.119574
0.29	4.005019	4.172013	4.347829	4.533039	4.728153	4.933862
0.30	3.876257	4.035217	4.202469	4.378506	4.563862	4.759112

Present Worth of a Series of Escalating Payments Compounded Annually

DISCOUNT-ESCALATION FACTORS FOR N = 13 YEARS

Discount Rate	Annual Escalation Rate					
	.05	.06	.07	.08	.09	.10
0.06	12.173937	13.000000	13.890931	14.852747	15.890747	17.009872
0.07	11.420135	12.181144	13.000000	13.882959	14.834741	15.860790
0.08	10.732747	11.434034	12.187882	13.000000	13.873964	14.816023
0.09	10.104909	10.751564	11.446958	12.195540	13.000000	13.865988
0.10	9.529762	10.127655	10.769736	11.460373	12.202319	13.000000
0.11	9.002319	9.555759	10.149790	10.788039	11.473105	12.209957
0.12	8.517930	9.031087	9.581337	10.171812	10.805602	11.485850
0.13	8.072157	8.548676	9.059301	9.606717	10.193678	10.822997
0.14	7.661287	8.104564	8.579001	9.087253	9.631481	10.214803
0.15	7.282104	7.694903	8.136470	8.609052	9.114616	9.656157
0.16	6.931475	7.316603	7.728099	8.168047	8.638485	9.141766
0.17	6.606863	6.966621	7.350717	7.760981	8.199263	8.667681
0.18	6.305870	6.642410	7.001372	7.384488	7.793386	8.229975
0.19	6.026304	6.341556	6.677545	7.035736	7.417735	7.825406
0.20	5.766313	6.062065	6.376932	6.712345	7.069716	7.450734
0.21	5.524240	5.801997	6.097507	6.412008	6.746790	7.103388
0.22	5.298429	5.559690	5.837358	6.132634	6.446665	6.780848
0.23	5.087553	5.333582	5.594850	5.872395	6.167380	6.481015
0.24	4.890390	5.122367	5.368484	5.629730	5.907097	6.201815
0.25	4.705760	4.924778	5.156911	5.403116	5.664281	5.941530
0.26	4.532685	4.739694	4.958926	5.191211	5.437451	5.698590
0.27	4.370210	4.566116	4.773385	4.992825	5.225242	5.471532
0.28	4.217531	4.403105	4.599311	4.806856	5.026481	5.259010
0.29	4.073852	4.249872	4.435797	4.632315	4.840064	5.059880
0.30	3.938515	4.105640	4.282035	4.468300	4.665085	4.873096

260   Life Cycle Costing: A Practical Guide for Energy Managers

Present Worth of a Series of Escalating Payments Compounded Annually

DISCOUNT-ESCALATION FACTORS FOR N = 14 YEARS

Discount Rate	Annual Escalation Rate					
	.05	.06	.07	.08	.09	.10
0.06	13.049719	14.000000	15.031332	16.151810	17.368759	18.689438
0.07	12.187997	13.058001	14.000000	15.022073	16.130707	17.333511
0.08	11.406832	12.203816	13.065778	14.000000	15.011571	16.108887
0.09	10.697409	11.428144	12.218582	13.074526	14.000000	15.002281
0.10	10.051144	10.723035	11.448748	12.233854	13.082315	14.000000
0.11	9.461651	10.080280	10.747997	11.469475	12.248382	13.091028
0.12	8.923064	9.493716	10.108955	10.772821	11.489394	12.262913
0.13	8.429885	8.957172	9.525186	10.137409	10.797463	11.509127
0.14	7.977502	8.465656	8.990821	9.556360	10.165187	10.821300
0.15	7.561925	8.014434	8.500892	9.024168	9.586898	10.192859
0.16	7.179352	7.599658	8.050919	8.535771	9.056853	9.617196
0.17	6.826674	7.217627	7.636985	8.087070	8.570262	9.089279
0.18	6.500991	6.865222	7.255486	7.673949	8.122712	8.604215
0.19	6.199681	6.539536	6.903340	7.292939	7.710361	8.157946
0.20	5.920522	6.238159	6.577765	6.941113	7.329990	7.746509
0.21	5.661536	5.958776	6.276313	6.615686	6.978518	7.366721
0.22	5.420781	5.699407	5.996702	6.314144	6.653172	7.015520
0.23	5.196692	5.458208	5.736985	6.034300	6.351584	6.690344
0.24	4.987833	5.233636	5.495387	5.774284	6.071558	6.388708
0.25	4.792837	5.024213	5.270316	5.532296	5.811251	6.108545
0.26	4.610572	4.828633	5.060361	5.306755	5.568909	5.847977
0.27	4.439936	4.645738	4.864193	5.096263	5.342925	5.605268
0.28	4.280006	4.474446	4.680676	4.899538	5.141927	5.378838
0.29	4.129877	4.313849	4.508762	4.715429	4.934625	5.167339
0.30	3.988800	4.163062	4.347524	4.542898	4.749956	4.969543

Present Worth of a Series of Escalating Payments Compounded Annually

DISCOUNT-ESCALATION FACTORS FOR N = 15 YEARS

Discount Rate	Annual Escalation Rate					
	.05	.06	.07	.08	.09	.10
0.06	13.917246	15.000000	16.182541	17.475388	18.888596	20.432404
0.07	12.941504	13.926670	15.000000	16.171783	17.450867	18.847534
0.08	12.062191	12.959342	13.935541	15.000000	16.159775	17.425629
0.09	11.268166	12.086102	12.976050	13.945451	15.000000	16.149063
0.10	10.548826	11.296765	12.109244	12.993276	13.954313	15.000000
0.11	9.896154	10.581173	11.324649	12.132495	13.009686	13.964153
0.12	9.302876	9.931562	10.613020	11.352365	12.154870	13.026094
0.13	8.762289	9.340363	9.966333	10.644620	11.379877	12.177039
0.14	8.268751	8.801409	9.377355	10.000779	10.675486	11.406518
0.15	7.817414	8.308957	8.839962	9.414018	10.034537	10.706226
0.16	7.403724	7.858313	8.348693	8.878136	9.449974	10.068035
0.17	7.023940	7.445033	7.898787	8.388077	8.915893	9.485654
0.18	6.674616	7.065373	7.485910	7.938881	8.426921	8.953083
0.19	6.352662	6.715890	7.106367	7.526368	7.978395	8.465335
0.20	6.055454	6.393708	6.756842	7.147005	7.566406	8.017635
0.21	5.780678	6.096120	6.434430	6.797482	7.187263	7.606115
0.22	5.526084	5.820799	6.136456	6.474826	6.837673	7.227109
0.23	5.289858	5.565610	5.860632	6.176462	6.514822	6.877548
0.24	5.070344	5.328754	5.604892	5.900186	6.216125	6.554501
0.25	4.865983	5.108534	5.367391	5.643908	5.939407	6.255518
0.26	4.675478	4.903456	5.146501	5.405793	5.682630	5.978393
0.27	4.497583	4.712193	4.940700	5.184227	5.443930	5.721101
0.28	4.331255	4.533525	4.748691	4.977738	5.221720	5.481814
0.29	4.175480	4.366420	4.569283	4.785014	5.014525	5.258970
0.30	4.029415	4.209883	4.401426	4.604873	4.821118	5.051153

Present Worth of a Series of Escalating Payments Compounded Annually

DISCOUNT-ESCALATION FACTORS FOR N = 16 YEARS

Discount Rate	Annual Escalation Rate					
	.05	.06	.07	.08	.09	.10
0.06	14.776581	16.000000	17.344574	18.823944	20.451431	22.241135
0.07	13.680928	14.787214	16.000000	17.332321	18.795746	20.403976
0.08	12.699345	13.700878	14.797251	16.000000	17.318588	18.766800
0.09	11.817976	12.725952	13.719622	14.808386	16.000000	17.306442
0.10	11.023886	11.849632	12.751723	13.738888	14.818386	16.000000
0.11	10.307171	11.059504	11.880520	12.777596	13.757276	14.829416
0.12	9.658952	10.345952	11.094582	11.911214	12.802521	13.775649
0.13	9.071159	9.699817	10.384058	11.129387	11.941672	12.827219
0.14	8.537008	9.113600	9.740153	10.421807	11.263402	11.971203
0.15	8.050687	8.580431	9.155445	9.780137	10.458822	11.197274
0.16	7.606818	8.094667	8.623364	9.196888	9.819372	10.495555
0.17	7.200974	7.651059	8.138213	8.665929	9.237891	9.858314
0.18	6.829112	7.245172	7.694855	8.181360	8.707926	9.278298
0.19	6.487644	6.872977	7.288920	7.738219	8.223905	8.749475
0.20	6.173521	6.531111	6.916518	7.332309	7.781150	8.266168
0.21	5.884065	6.216438	6.574254	6.959745	7.375308	7.823745
0.22	5.616714	5.926271	6.259027	6.617068	7.002516	7.417886
0.23	5.369390	5.658167	5.968194	6.301287	6.659481	7.044967
0.24	5.140212	5.410065	5.699384	6.009843	6.343204	6.701573
0.25	4.927424	5.180038	5.450486	5.740339	6.051161	6.384854
0.26	4.729566	4.966401	5.219650	5.490683	5.781007	6.092249
0.27	4.545245	4.767660	5.005158	5.259030	5.530619	5.821430
0.28	4.373296	4.582451	4.805547	5.043719	5.298185	5.570311
0.29	4.212599	4.409616	4.619483	4.843270	5.082038	5.337105
0.30	4.062220	4.248059	4.445792	4.656359	4.880785	5.120207

Present Worth of a Series of Escalating Payments Compounded Annually

DISCOUNT-ESCALATION FACTORS FOR N = 17 YEARS

Discount Rate	Annual Escalation Rate					
	.05	.06	.07	.08	.09	.10
0.06	15.627816	17.000000	18.517624	20.197876	22.058517	24.118134
0.07	14.406532	15.639714	17.000000	18.503571	20.165741	22.004089
0.08	13.318802	14.428679	15.650982	17.000000	18.488220	20.132797
0.09	12.347609	13.348191	14.449548	15.663403	17.000000	18.474304
0.10	11.477351	12.382395	13.376683	14.470946	15.674597	17.000000
0.11	10.695970	11.516288	12.416359	13.405260	14.491394	15.686883
0.12	9.992771	10.738141	11.554645	12.450102	13.432822	14.511818
0.13	9.358163	10.037003	10.779602	11.592706	12.483582	13.460139
0.14	8.784086	9.403883	10.080674	10.820676	11.629918	12.516074
0.15	8.263676	8.830659	9.448982	10.123970	10.860970	11.666971
0.16	7.790654	8.310648	8.876723	9.493658	10.166479	10.900961
0.17	7.359850	7.837715	8.357175	8.922408	9.537873	10.208678
0.18	6.966588	7.406685	7.884321	8.403291	8.967499	9.581465
0.19	6.606746	7.012904	7.453066	7.930488	8.448785	9.012128
0.20	6.276829	6.652482	7.058896	7.499081	7.976209	8.493988
0.21	5.973781	6.321840	6.697899	7.104576	7.544703	8.021591
0.22	5.694715	6.017911	6.366527	6.742988	7.149793	7.589898
0.23	5.437284	5.737931	6.061765	6.410889	6.787673	7.194690
0.24	5.199375	5.479571	5.780921	6.105350	6.454910	6.832043
0.25	4.979035	5.240674	5.521616	5.823656	6.148609	6.498670
0.26	4.774638	5.019356	5.281770	5.563445	5.866111	6.191647
0.27	4.584650	4.813956	5.059466	5.322643	5.605021	5.908329
0.28	4.407782	4.622967	4.853076	5.099392	5.363300	5.646361
0.29	4.242811	4.445112	4.661122	4.892044	5.139083	5.403731
0.30	4.088716	4.279188	4.482309	4.699131	4.930813	5.178638

264 Life Cycle Costing: A Practical Guide for Energy Managers

Present Worth of a Series of Escalating Payments Compounded Annually

DISCOUNT-ESCALATION FACTORS FOR N = 18 YEARS

Discount Rate	Annual Escalation Rate					
	.05	.06	.07	.08	.09	.10
0.06	16.471008	18.000000	19.701706	21.597778	23.711090	26.065933
0.07	15.118572	16.484253	18.000000	19.685944	21.561371	23.649033
0.08	13.921048	15.143003	16.496796	18.000000	19.668549	21.524109
0.09	12.857807	13.953305	15.166082	16.510559	18.000000	19.652985
0.10	11.910205	12.895785	13.984599	15.189694	16.523026	18.000000
0.11	11.063752	11.952495	12.932888	14.015963	15.212285	16.536621
0.12	10.305727	11.109321	11.994169	12.969746	14.046240	15.234840
0.13	9.624847	10.353301	11.154145	12.035522	13.006309	14.076256
0.14	9.011659	9.673796	10.400288	11.198551	12.075973	13.041827
0.15	8.458143	9.061305	9.722098	10.446876	11.242136	12.116246
0.16	7.957057	8.508009	9.110427	9.769960	10.492640	11.285398
0.17	7.502432	8.006823	8.557423	9.159158	9.817344	10.538080
0.18	7.088918	7.551773	8.056126	8.606414	9.207274	9.864078
0.19	6.711836	7.137545	7.600657	8.104983	8.654766	9.254914
0.20	6.367223	6.759694	7.185849	7.649177	8.153387	8.702825
0.21	6.051633	6.414176	6.807238	7.233846	7.697299	8.201451
0.22	5.761847	6.097533	6.460810	6.854458	7.281376	7.744991
0.23	5.495242	5.806672	6.143164	6.507125	6.901275	7.328588
0.24	5.249473	5.538989	5.851279	6.188534	6.553103	6.947781
0.25	5.022389	5.292093	5.582503	5.895642	6.233584	6.598827
0.26	4.812200	5.063905	5.334522	5.625813	5.939733	6.278422
0.27	4.617229	4.852596	5.105222	5.376740	5.668879	5.983595
0.28	4.436071	4.656520	4.892807	5.146365	5.418749	5.711718
0.29	4.267403	4.474279	4.695660	4.932877	5.187284	5.460545
0.30	4.110116	4.304570	4.512365	4.734666	4.972759	5.228079

Present Worth of a Series of Escalating Payments Compounded Annually

DISCOUNT-ESCALATION FACTORS FOR N = 19 YEARS

Discount Rate	Annual Escalation Rate					
	.05	.06	.07	.08	.09	.10
0.06	17.306259	19.000000	20.896896	23.024094	25.410416	28.087219
0.07	15.817303	17.320892	19.000000	20.879227	22.983017	25.340134
0.08	14.506567	15.844100	17.334793	19.000000	20.859909	22.941193
0.09	13.349283	14.541765	15.869470	17.349960	19.000000	20.842453
0.10	12.323383	13.390506	14.575933	15.895371	17.363739	19.000000
0.11	11.411654	12.369054	13.430804	14.610157	15.920185	17.378708
0.12	10.599122	11.460615	12.414070	13.470829	14.643228	15.944954
0.13	9.872652	10.650006	11.508800	12.458747	13.510531	14.676015
0.14	9.221265	9.924767	10.700274	11.556540	12.502464	13.549130
0.15	8.635700	9.273899	9.976215	10.750126	11.603415	12.545987
0.16	8.107680	8.688356	9.325996	10.027208	10.799118	11.649950
0.17	7.630389	8.160031	8.740556	9.377697	10.077704	10.847775
0.18	7.197771	7.682107	8.211915	8.792323	9.428761	10.127531
0.19	6.804562	7.248569	7.733367	8.263350	8.843438	9.479340
0.20	6.446318	6.854398	7.299050	7.784263	8.314324	8.894258
0.21	6.119191	6.495066	6.903926	7.349228	7.834762	8.364961
0.22	5.819625	6.166713	6.543501	6.953135	7.398938	7.884829
0.23	5.544719	5.865911	6.213973	6.591625	7.001946	7.448337
0.24	5.291893	5.589782	5.911992	6.260984	6.639419	7.050452
0.25	5.058806	5.335696	5.634623	5.957838	6.307682	6.686966
0.26	4.843500	5.101382	5.379319	5.679272	6.003422	6.354178
0.27	4.644164	4.884847	5.143772	5.422742	5.723685	6.048787
0.28	4.459277	4.684305	4.926020	5.185999	5.465967	5.767883
0.29	4.287419	4.498245	4.724307	4.967063	5.228012	5.508990
0.30	4.127401	4.325267	4.537104	4.764187	5.007930	5.269914

Present Worth of a Series of Escalating Payments Compounded Annually

DISCOUNT-ESCALATION FACTORS FOR N = 20 YEARS

Discount Rate	Annual Escalation Rate					
	.05	.06	.07	.08	.09	.10
0.06	18.133621	20.000000	22.103424	24.477341	27.157867	30.184814
0.07	16.502960	18.149719	20.000000	22.083740	24.431290	27.078613
0.08	15.075819	16.532211	18.165024	20.000000	22.062271	24.384445
0.09	13.822721	15.114027	16.559937	18.181671	20.000000	22.042847
0.10	12.717781	13.867238	15.151142	16.588211	18.196808	20.000000
0.11	11.740750	12.766849	13.910778	15.188293	16.615326	18.213211
0.12	10.874181	11.793090	12.815228	13.954018	15.224222	16.642380
0.13	10.102913	10.928331	11.844624	12.863245	13.996905	15.259852
0.14	9.414323	10.158126	10.981841	11.895686	12.910251	14.038637
0.15	8.797817	9.469856	10.212655	11.034917	11.945845	12.957045
0.16	8.244020	8.853156	9.524842	10.266715	11.087101	11.995647
0.17	7.745223	8.298836	8.908035	9.579425	10.320264	11.138941
0.18	7.294631	7.799186	8.353181	8.962478	9.633354	10.373123
0.19	6.886379	7.347465	7.852694	8.407076	9.016256	9.686791
0.20	6.515526	6.938053	7.399987	7.905839	8.460508	9.069737
0.21	6.177816	6.565928	6.989428	7.452213	7.958591	8.513605
0.22	5.869351	6.226819	6.616025	7.040490	7.503973	8.010912
0.23	5.586954	5.916964	6.275573	6.665820	7.091159	7.555427
0.24	5.327814	5.633201	5.964381	6.324086	6.715292	7.141531
0.25	5.089396	5.372672	5.679236	6.011576	6.372296	6.764528
0.26	4.869584	5.132911	5.417361	5.725093	6.058517	6.420316
0.27	4.666433	4.911765	5.176250	5.461863	5.770723	6.105252
0.28	4.478313	4.707315	4.953783	5.219440	5.506177	5.816151
0.29	4.303712	4.517939	4.748069	4.995684	5.262425	5.550301
0.30	4.141362	4.342142	4.557465	4.788713	5.037419	5.305312

Present Worth of a Series of Escalating Payments Compounded Annually

DISCOUNT-ESCALATION FACTORS FOR N = 25 YEARS

Discount Rate	Annual Escalation Rate					
	.05	.06	.07	.08	.09	.10
0.06	22.154831	25.000000	28.308899	32.165253	36.665466	41.921478
0.07	19.743790	22.179077	25.000000	28.277069	32.088928	36.530167
0.08	17.693497	19.786270	22.202301	25.000000	28.243073	32.011902
0.09	15.941714	17.747574	19.826904	22.227112	25.000000	28.211899
0.10	14.436612	16.003113	17.800247	19.868027	22.249954	25.000000
0.11	13.137866	14.502648	16.063324	17.852875	19.907623	22.274292
0.12	12.012110	13.206630	14.567844	16.123108	17.903931	19.947098
0.13	11.031583	12.081730	13.274472	14.632588	16.182388	17.954605
0.14	10.173676	11.101089	12.150617	13.341736	14.696076	16.240265
0.15	9.419918	10.242241	11.169856	12.219010	13.407931	14.759292
0.16	8.754628	9.487017	10.310243	11.238118	12.286388	13.473707
0.17	8.165100	8.819980	9.553672	10.377841	11.305828	12.353399
0.18	7.640522	8.228430	8.884878	9.619890	10.444737	11.372787
0.19	7.171824	7.701640	8.291338	8.949347	9.685426	10.511131
0.20	6.751500	7.230802	7.762498	8.353923	9.013366	9.750687
0.21	6.373233	6.808322	7.289540	7.823112	8.416147	9.077086
0.22	6.031450	6.427856	6.864894	7.348015	7.883290	8.477970
0.23	5.721639	6.083976	6.482291	6.921221	7.406138	7.943212
0.24	5.439904	5.772174	6.136369	6.536545	6.977273	7.464020
0.25	5.182830	5.488499	5.822557	6.188597	6.590548	7.033149
0.26	4.947592	5.229600	5.536984	5.872822	6.240630	6.644423
0.27	4.731661	4.992602	5.276248	5.585337	5.922914	6.292513
0.28	4.532938	4.775002	5.037500	5.322799	5.633556	5.972860
0.29	4.349524	4.574699	4.818263	5.082335	5.369199	5.681642
0.30	4.179843	4.389812	4.616408	4.861465	5.127053	5.415556

Present Worth of a Series of Escalating Payments Compounded Annually

DISCOUNT-ESCALATION FACTORS FOR N = 30 YEARS

Discount Rate	Annual Escalation Rate					
	.05	.06	.07	.08	.09	.10
0.06	25.989899	30.000000	34.812729	40.606293	47.596725	56.046082
0.07	22.692841	26.023636	30.000000	34.765274	40.489456	47.383087
0.08	19.967239	22.749985	26.056076	30.000000	34.715256	40.372177
0.09	17.699402	20.038132	22.804932	26.090378	30.000000	34.669113
0.10	15.798732	17.777908	20.107300	22.860321	26.122162	30.000000
0.11	14.196053	15.881169	17.854996	20.176315	22.913818	26.155685
0.12	12.836191	14.279993	15.962653	17.931549	20.243454	22.967102
0.13	11.674885	12.919486	14.362946	16.043625	18.007507	20.310135
0.14	10.677021	11.756478	13.002002	14.445253	16.123138	18.081818
0.15	9.814664	10.756134	11.837328	13.084011	14.526378	16.202362
0.16	9.064901	9.890878	10.834711	11.917679	13.164935	14.607058
0.17	8.409522	9.138076	9.966700	10.912924	11.997480	13.245506
0.18	7.833486	8.479517	9.210846	10.042119	10.990429	12.076521
0.19	7.324487	7.900255	8.549147	9.283236	10.116879	11.067464
0.20	6.872533	7.388243	7.966828	8.618512	9.355217	10.191415
0.21	6.469390	6.933384	7.451824	8.033224	8.687572	9.426965
0.22	6.107997	6.527399	6.994043	7.515202	8.099232	8.756285
0.23	5.782696	6.163363	6.585276	7.054516	7.578281	8.165045
0.24	5.488702	5.835612	6.218651	6.643029	7.114770	7.641182
0.25	5.221905	5.539289	5.888426	6.273827	6.700584	7.174908
0.26	4.978941	5.270343	5.589815	5.941172	6.328860	6.758073
0.27	4.756858	5.025345	5.318698	5.640250	5.993789	6.383796
0.28	4.553228	4.801364	5.071674	5.366998	5.690596	6.046313
0.29	4.365892	4.595963	4.845822	5.117975	5.415186	5.740853
0.30	4.193070	4.406992	4.638674	4.890255	5.164197	5.463374

Solutions

SOLUTIONS TO PROBLEMS

CHAPTER 2

2-1 Find present value of an annuity:
$1,000 (UPW: 5, 8%) =
$1,000 (3.99271) = $3,992.71

2-2 a. 30,000 × (UCR, 20, 12%)
30,000 × .13388 = $4,016.40
b. 30,000 × (UCR × 240, 1%)

2-3 First calculate the present value of all obligations at 12% per annum:
a. Mortgage:
PV = $2,400 (UPW: 20, 12%)
= $2,400 (7.46944)
= $17,926.65
b. Auto loan:
PV = $75 (UPW; 12, 2%)
= $75 (10.57511)
= $793.13
c. Debts due in 2 years:
PV = $6,000 (SPW; 2, 12%)
= $6,000 (0.79719)
= $4,783.14
d. Current debt:
PV = $700 (SPW; 0, 12%)
= $700 (1.0)
= $700

Present value of total indebtedness:
Mortgage	$17,926.65
Auto loan	793.13
Other debt	4,783.14
Current debt	700.00
Total	$24,202.92

To calculate the annual amount necessary to retire the debt, use Capital Recovery Factor:
$24,202.92 (UCR; 20, 12%) =
$24,202.92 (0.13388) = $3,240.29 ANSWER

2-4 a. Calculate future value of a single sum:
$5,000 (SCA; 4, 14%) =
$5,000 (1.68896)   = $8,444.80
b. $5,000 (SCA; 8, 7%) =
$5,000 (1.71818)   = $8,590.90

2-5 First find present value of $8,000, 16 periods from now at a discount rate of 2.00% per period:
$8,000 (SPW; 16, 2.00%) =
$8,000 (.72845)   = $5,827.60
This is the present deposit needed to accumulate to $8,000 in 4 years. Subtracting from $12,000 yields the amount available for current use:
$12,000 − $5,827.60 = $6,172.40

2-6 Amount saved annually = $25,000 − $15,000 = $10,000

Amount saved quarterly = $\frac{\$10{,}000}{4}$ = $2,500

Find future value of an annuity of $2,500 per period, for 20 periods compounding at a rate of 2.00% per period:
$2,500 (UCA; 20, 2.00%) =
$2,500 (24.29662)   = $60,741.55

2-7 Use the (USF) factor:
$5,000 × (USF; 5, 10%) =
$5,000 × (.16380)   = $819

2-8 Mary must make a total of 10 with her first payment due today. Use (USF) factor:
$3,000 × (USF; 10, 5%) =
$3,000 × (.07951)   = $238.53

2-9 a. Find present value of an annuity:
$5,000 (UPW; 20, 5%) =
$5,000 (12.46210) = $62,310.50
b. Find capital recovery factor:
$250,000 (UCR; 20, 5%) =
$250,000 (0.08024) = $20,060
This amounts to about 4 awards of $5,000 each.

2-10 Find present value of an annuity:
$100 (UPW; 10, 8%) =
$100 (6.71008)   = $671.01

2-11 Go to 5% tables and find the first Single Payment Compound Amount Factor which is equal to or greater than 2.

(SCA; 15, 5%) = 2.0789
Answer: 15 years.
(SCA; 12, 6%) = 2.01218
Answer: 12 years.

To find out how long it will take money to triple, repeat the above process, looking for a factor equal to or greater than 3.

(SCA; 23, 5%) = 3.0714
Answer: 23 years.
(SCA; 19, 6%) = 3.02557
Answer: 19 years.

2-12 $159 = $2,000 (USF; 10, ?)

$$(USF; 10, ?) = \frac{\$159}{\$2,000} = 0.0795$$

(USF; 10, 5%) = 0.0795
Answer: 5%

2-13 Calculate present value of a single sum:
$10,000 (SPW; 15, 10%) =
$10,000 (0.23939)   = $2,393.90

2-14 Calculate future value of $3,000:
$3,000 (SCA; 4, 5%)
$3,000 × 1.21550 = $3,646.50
Interest = $3,646.50 − $3,000 = $646.50

2-15 a. Calculate future value of a single sum:
$1,000 (SCA; 5, 6%) =
$1,000 (1.33822)   = $1,338.22
b. $1,000 (SCA; 3, 6%) =
$1,000 (1.19101)   = $1,191.01
c. $1,000 (SCA; 5, 7%) =
$1,000 (1.40255)   = $1,402.55
d. $1,000 (SCA; 10, 3%) =
$1,000 (1.34391)   = $1,343.91

CHAPTER 3

3-1 Present worth of $3,000/yr for 20 yrs
$3,000 × (UPW, 10%, 20)
$3,000 × 8.51355 = $ 25,541
Present worth of $2,000/yr for 15 yrs
$2,000 × (UPW; 10%, 15)
$2,000 × 7.60606 = 15,212
Present worth of $7,000/yr for 10 yrs
$7,000 × (UPW; 10%, 10)
$7,000 × 6.14455 = 43,012
Salvage value
$10,000 × (SPW; 10%, 20)
$10,000 × .14865 = 1,487
Total present worth of savings and
 salvage = $ 85,252

The present worth of both savings and salvage does not equal the total amount invested, so it is not advisable to invest in this new machinery.

3-2 Use uniform annual costing method.
Initial cost
 $2,000,000 × (UCR; ∞, 12)
 $2,000,000 × .12 = $240,000
Annual operating cost   200,000
Total annual cost $440,000
Expected annual savings $500,000
Expected annual cost   440,000
Net savings per year   = $ 60,000 Answer

3-3 a. Initial cost $240,000
b. Operation cost
 3600 hours × $8 = $28,800
 $28,800 × (UPW; 3, 10%) =
 $28,800 × 2.48684 =   71,621
c. Repair cost
 60 × 10   = 600 hours
 600 × $10 = 6,000
 6,000 × (UPW; 3, 10%)
 6,000 × 2.48684 =   14,921

d. Power cost
  3,000 hours × 10KW × .03 = 900
  900 × (UPW; 3, 10%)
  900 × 2.48684   = $ 2,238
e. Salvage
  100,000 × (SPW; 3, 10%)
  100,000 × .75132   =   (75,132)
        LCC $253,648 Answer

CHAPTER 4

4-1 At escalation rates of 8%, 9%, and 10% the break–even points are 2.2 years or less. The probability of having an escalation rate of 8% or higher is .40 + .20 + .10 = .70. Since this is less than the required 90% probability level the company should reject the project.

4-2 Classification of some of the factors will depend upon circumstances. For example, if the plot has already been purchased, its area is a constant. If it has not been purchased, and a variety of options are available, it is a parameter. Here is one possible grouping:

Constants	*Variables*	*Parameters*
Size of classrooms	Energy inflation rate	Shape of building
Number of classrooms	Costs of materials	HVAC system
Area of plot		Materials used
Bathroom areas		
Discount rate		

4-3 At $11 an hour:
  .15 × 1 = .15
  .50 × 4 = 2.00
  .25 × 7 = 1.75
  .10 × 10 = 1.00
           ─────
           4.90 hours

Expected annual labor cost = 3.5 × 4.9 × $11 = $188.65
  At $12 an hour:
  .20 × 1 = .20
  .55 × 4 = 2.20
  .15 × 7 = 1.05
  .10 × 10 = 1.00
           ─────
           4.45 hours

Expected annual labor cost = 3.5 × 4.45 × $12 = $186.90
Answer: Optimal wage rate = $12 an hour.

CHAPTER 6

6-1

Year	Average Annual Capital Costs	Average Annual Maintenance/ Repair Costs	Total Average Annual Costs
1	32,000	0	32,000
2	25,954	1,395	27,343
3	23,519	2,145	25,664
4	21,012	2,917	23,929
5	19,416	2,820	23,236
6	18,283	4,983	23,266
7	17,422	5,888	23,310
8	16,735	6,770	23,505

Service life is 5 years.

CHAPTER 7

7-1   a. Straight-line:
      42,000 ÷ 6 = $7,000 per year
   b. SYD:
      6/21 × 42,000 = 12,000
      5/21 × 42,000 = 10,000
      4/21 × 42,000 = 8,000
      3/21 × 42,000 = 6,000
      2/21 × 42,000 = 4,000
      1/21 × 42,000 = 2,000
   c. DDB:
      1/3 × 45,000 = 15,000
      1/3 × 30,000 = 10,000
      1/3 × 20,000 = 6,667
      1/3 × 13,333 = 4,444
      1/3 × 8,889 = 2,963
     Year 6: 42,000 − 39,074 = 2,926

7-2

Present Worth	System A	System B
Initial cost	120,000	90,000
Operating & maintenance costs	45,249	56,561
Property tax	21,720	16,290
Depreciation tax benefit	(15,083)	(11,312)
Present Worth Totals	171,886	151,539

System B
Operating & maintenance costs
$\$10,000 \times .60 \times 9.42691 = 56,561$
Property tax
$$\frac{90,000}{1,000} \times .40 \times \$80$$
$= 2,880 \times .60 \times 9.42691 = 16,290$
Depreciation tax benefit
$$\frac{90,000}{30} = \$3,000$$
$3,000 \times .40 \times 9.42691 = 11,312$

Answer: System B is superior.

CHAPTER 8

8-1 Rental
$\$1,400 \times 20$ miles = $28,000 per year
Purchase
  a. $17,000 \times$ (UCR; 10, 10%)
    $17,000 \times .16275 =$ 2,767
  b. Labor, etc.   20,000
                        $22,767

Answer: Savings = $28,000 - 22,767 = \$5,233$ a year.

8-2   a. Costs: Maintenance, license, insurance, storage, supplies, replacements, inspection.
     Life of Truck
     Salvage value of Truck
     Discount rate

Note: Since labor would have to be supplied equally for a

rental or owned truck, it is not included in the analysis. This applies to fuel also unless there is reason to believe that the fuel consumption would differ.

b. The uniform annual cost method is preferable since the borough manager would probably want to know the difference between average annual costs of the alternatives.

8-3 Lease
    Annual costs:

Mileage .075 × 25,000	1,875
Fixed charge $200 × 12	2,400
	4,275

    Annual tax benefits:
      Business use .75 × 4,275 = 3,206

Tax deduction .25 × 3,206	(802)
Total annual cost	3,473

    Present worth:
      3,473 × (UPW; 3, 9%)

3,473 × 2.53128 =	8,791

Buy
    Annual costs:

Operating & maintenance .08 × 25,000	2,000
Insurance and taxes	574
	2,574

    Annual tax benefits:

Operating & maintenance		
2,000 × .75 × .25 =	375	
Insurance 550 × .75 × .25 =	103	
Local taxes 24 × .25 =	6	
Depreciation		
(6,000 − 2,000) × 1/3 = 1,333		
1,333 × .75 × .25 =	250	
Total tax benefits =		(734)
Annual costs less tax benefits		1,840

    Present worth:

Initial cost	6,000
Salvage 2,000 × (SPW; 3, 9%)	
2,000 × .77219 =	(1,544)

Annual costs less tax benefits
1,840 × (UPW; 3, 9%)
1,840 × 2.53128 =   4,658
Investment tax credit
6,000 × 75% × .10 × 1/3 =   (150)
                                                8,964

Answer: Leasing is more attractive.

8-4 Lease
Rental $40,000
Costs   6,000
Total $46,000

After-tax average annual cost = 46,000 × .60 = $27,600.
Present worth:
27,600 × (UPW; 10, 6%)
27,600 × 7.36004 =   203,137

Buy
Initial cost   300,000
Present Worth Salvage
180,000 × (SPW; 10, 6%)
180,000 × .55840 =   (100,512)
Present Worth Annual Costs
9,000 × .60 × (UPW; 10, 6%)
9,000 × .60 × 7.36004 =   39,744
Present Worth Depreciation tax benefit
120,000 ÷ .10 = 12,000
12,000 × .4 × (UPW; 10, 6%)
12,000 × .4 × 7.36004 =   (35,328)
                                                203,904

Answer: The lease is more attractive.

8-5 Lease
$275,000 × .60 = $165,000
Payments: At t_0 275,000
$t_1 - t_4$ 165,000 × (UPW; 4, 6%)
            165,000 × 3.46508   571,738
t_5 tax benefit of 110,000
     110,000 × (SPW; 5, 6%)

Solutions to Problems

	110,000 × .74726	(82,199)
		764,539

Buy
Initial cost $1,071,428

Investment Tax Credit
   1,071,428 × .10 × 2/3 =   (71,428)

Present Worth Salvage 71,428 × (SPW; 5, 6%)
                          71,428 × .74726 =   (53,375)

Present Worth Maintenance
   40,000 × .6 × (UPW; 5, 6%)
   40,000 × .6 × 4.21233 =   101,096

Depreciation tax benefit
   5/15 × 1,000,000 × .4
     × (SPW; 1, 6%) =   125,787
   4/15 × 1,000,000 × .4
     × (SPW; 2, 6%) =   94,933
   3/15 × 1,000,000 × .4
     × (SPW; 3, 6%) =   67,170
   2/15 × 1,000,000 × .4
     × (SPW; 4, 6%) =   42,245
   1/15 × 1,000,000 × .4
     × (SPW; 5, 6%) =   19,927

                                                 (350,062)
                                                697,659

Answer: Purchase is more attractive.

CHAPTER 9

9-1   Plan A

Initial cost		= $10,000
Present worth of annual costs		
5500 × (UPW; 5, 10%)		
5500 × 3.79077		= 20,849
Present worth of salvage value		
3000 × (SPW; 5, 10%)		
3000 × .62092		= (1,863)
Total Present Worth		= $28,986

Uniform annual cost
28,986 × (UCR; 5, 10%)
28,986 × .26380 = $ 7,647

Plan B
*Initial cost 50,000 + 10,000 − 15,000 = $45,000
Present worth of annual operating &
maintenance cost
2500 × (UPW; 25, 10%)
2500 × 9.07703 = $22,693
Present worth of salvage value
3000 × (SPW; 25, 10%)
3000 × .09230 = (277)
Total present worth = $67,416
Uniform annual cost
67,416 × (UCR; 25, 10%)
67,416 × .11017 = $ 7,427

9-2 We can solve this problem in the usual way by finding the Present Worth of each bus and subtracting. However, we can shorten the work since we do not need the P.W. of each, but only the difference between them. By the usual way we have to obtain the SPW of each of the 5 annual costs for each bus. However, by subtracting the annual costs of the new bus from those of the old bus and treating the $900 difference for each year as a uniform annual amount, we have to make just the one UPW calculation.

a. Old bus: Initial cost $10,000
              Annual costs 900 × (UPW; 5, 8%)
                           900 × 3.99271   3,593
                                              $13,593
   New bus: Initial cost $20,000
              Salvage 10,000 × (SPW; 5, 8%)
                      10,000 × .68058   (6,806)
                                              $13,194

---

*Think of it this way: The old equipment, worth $10,000, is financed by $10,000 debt. The $15,000 trade-in will pay off the debt and leave a balance of $5,000 to apply against the $50,000 price.

Solutions to Problems

Old bus	$13,593
New bus	13,194
Difference in favor of new bus	399

b. Use same procedure for years 4, 6, etc. to determine the replacement year when savings are maximized.

9-3 Alternative B

P. W. Initial cost	$10,000
P. W. Operating costs 2,000 × (UPW; 8, 8%) 2,000 × 5.74664	11,493
P. W. Replacement 12,000 × (SPW; 4, 8%) 12,000 × .73503	8,820
P. W. Salvage – 4 years 3,000 × (SPW; 4, 8%) 3,000 × .73503	(2,205)
P. W. Salvage – 8 years 3,000 × (SPW; 8, 8%) 3,000 × .54027	(1,621)
LCC =	$26,487

Difference in favor of A is $26,487 − $20,919 = $5,568, Answer.

9-4 Old bus:

Initial cost	$10,000
Annual costs 3,593 × .60	2,156
Depreciation (10,000 ÷ 5) 2,000 (.4) (UPW; 5, 8%)	(3,194)
	$ 8,962

New bus:

Initial cost less P. W. of Salvage	$13,194
Depreciation (10,000 ÷ 5) 2,000 (.4) (UPW; 5, 8%)	(3,194)
	$10,000

New bus	$10,000
Old bus	8,962
Difference in favor of old bus	$ 1,038 Answer

9-5 a. Old

Initial Cost	$ 6,000
Depreciation Tax Benefits .46 (600) (UPW; 10, 12%)	(1,559)
	$ 4,441

282   Life Cycle Costing: A Practical Guide for Energy Managers

New
   Initial Cost $14,000
   Investment Tax Credit   (1,400)
   Depreciation Tax Benefits
     .46 (1,300) (UPW; 10, 12%)   (3,379)
   Salvage 1,000 (SPW; 10, 12%)   (322)
   Savings 3,000 (.54) (UPW; 10, 12%)   (9,153)
                                            $ (254)

   Total Savings = 4,441 + 254 = $4,695
b. Old
   Initial Cost
     $2,000 + .46 (4,000) $ 3,840
   Depreciation Tax Benefits .46 (600)
     (UPW; 10, 12%)   (1,559)
                                            $ 2,281

   Total Savings = 2,281 + 254 = $2,535

CHAPTER 10

10-1   40,000 gals. × $.50   = $ 20,000
       $20,000 × 10.54882 = 210,976 Answer

10-2   To get life: Longest life is 36,000 hours or approximately 4 years.

Power Costs — Expressed as a present worth value
a. Incandescent

$$\frac{150 \text{ watts} \times 8760 \text{ hr} \times \$.03/\text{KWH}}{1000} = \$39.42$$

7% escalation, 7% discount factor

P.W. Power Cost =

$$\$39.42 \left[\frac{1.07}{1.07} + \frac{(1.07)^2}{(1.07)^2} + \ldots + \frac{(1.07)^4}{(1.07)^4} \right]$$

= 39.42 × (4) = $157.68

b. Circleline Fluorescent

$$\frac{60 \text{ watts} \times 8760 \times .03}{1000} = \$15.77$$

P.W. Power Cost = $15.77 × 4   = 63.07

c. 40 watt, 4 foot strip Fluorescent

$$\frac{40 \text{ watt} \times 8760 \times .03}{1000} = \$10.50$$

P.W. Power Cost = 42.00

Lamp Replacement Costs

a. Incandescent

$$\text{No. replacement/yr} = \frac{8760 \text{ hr/yr}}{2500 \text{ hr}} = 3.5/\text{yr}$$

($1.18 + $.50) × 3.5 = $5.88/yr

P.W. Replacement Cost

$$= 5.88 \left[\frac{1.05}{1.07} + \frac{(1.05)^2}{(1.07)^2} + \frac{(1.05)^3}{(1.07)^3} + \frac{(1.05)^4}{(1.07)^4} \right]$$

= 5.88 [.9813 + .9629 + .9449 + .9270]
= 5.88 [3.816]
= $22.44

c. Circleline Fluorescent

$$\text{No. replacement/yr} = \frac{8760}{22,500} = .389$$

P. W. Replacement Cost =
($13.15 + .50) × .389 × 3.816 = $20.26

c. 40 watt, 4 foot Fluorescent

$$\text{No. replacement/yr} = \frac{8760}{36,000} = .243/\text{yr}$$

P. W. Replacement Cost
= (1.53 + .50) × .243 × 3.816
= $1.88

P. W. Cost per lighting fixture

	Incandescent	Circleline	40 Watt
Initial Cost	0	34.85	25.00
Power Cost	157.68	63.07	42.00
Replacement	22.44	20.26	1.88
Total P. W. Cost	$180.12	$118.18	$68.88

Uniform Annual Cost per lighting fixture
P.W. × (UCR, 4, 7%)

Incandescent	Circleline	40 Watt
$180.12 × .2952	$111.87 × .2952	$66.88 × .2952
$53.17	$34.88	$20.33

Conclusion: To change the 288 incandescent lighting fixtures to fluorescent strip lighting would save the Commonwealth $32,037 and 1,110,067 KWH over a 4-year period.

10-3 Solution of EP vs. FFBC problem

a. EP

Initial Cost	$14,100,000
Operating Costs	
1,150,693 × 11.89568	13,688,275
Maintenance Costs	
Annual	
$73,800 × 11.89568	877,902
7 years	
(1) $480,000 × $\frac{(1.08)^7}{(1.14)^7}$	328,757
(2) $480,000 × $\frac{(1.08)^{14}}{(1.14)^{14}}$	225,169
Total	$29,220,103

FFBC

Initial Cost	$12,500,000
Operating Costs	
798,278 × 11.89568	9,496,060
Maintenance Costs	
Annual: 50,000 × 11.89568	594,784
* 2 years: 636,768 × 5.44973	3,470,214
7 years: (1) 70,000 × .684911	47,944
(2) 70,000 × .4691022	32,837
Total	$26,141,839
Difference in favor of FFBC:	$ 3,078,264

b. EP

Initial Cost	$14,100,000
Operating Costs $13,688,275 × .60	8,212,965
Maintenance Costs	
877,902 × .60	526,741
328,757 × .60	197,254
225,169 × .60	135,101

---

* adj i = $\frac{(1 + .14)^2}{(1 + .08)} - 1 = .2033$

where n = 9

Solutions to Problems

Depreciation Tax Benefits
Average annual depreciation =
14,100,000 ÷ 20 = $705,000
705,000 × (UPW; 20, 14%) × .40 = $(1,876,723)
14,100,000 × .20 (Investment
    Tax Credit) (1,410,000)
                              Total $ 19,894,338

FFBC
Initial Cost $ 12,500,000
Operating Costs $9,496,060 × .60   5,697,636
Maintenance Costs
  594,784 × .60   356,870
  3,470,214 × .60   2,082,128
  80,781 × .60   48,469
Depreciation Tax Benefits
12,500,000 ÷ 20 = $625,000
625,000 × (UPW; 20, 14%) × .40 = (1,655,782)
12,500,000 × .10 (Investment
    Tax Credit) (1,250,000)
                              Total $ 17,779,321
Difference in favor of FFBC: $ 2,115,017

10-4 Life Cycle Cost Analysis — Short Form
Using Present Worth and Uniform Annual Cost Methods

Project Name: Telephone Installation   Life Cycle: 20 yrs
Location: Hometown, U.S.A.   Interest Rate: 8 %
Owner/Engineer: Bill Til
                                            Escalation Rate: 10 %

		Alternate No. 1	Alternate No. 2	Alternate No. 3
Initial Costs	1. Initial Costs			
	a. Base Cost			
	b. Interface & Auxiliary Costs			
	I.			
	II.			
	III.			
	IV.			
	V.			
	c. TOTAL INITIAL COST	12,600	8,568	
	d. Difference in Initial Cost	4,032		

		Alternate No. 1	Alternate No. 2	Alternate No. 3
Total Life Cycle Costs (Present Worth)	2. Operating Cost, P.W.			
	a. Fuel Cost			
	b. Operating Labor Cost			
	c. Maintenance Cost			
	d. Replacement Cost			
	e. Salvage Value ( )			
	f. Service Costs (24.38444) Discount-Escalation	69,130	121,800	
	g. _____			
	h. _____			
	i. TOTAL OPERATING COST	69,130	121,800	
	3. TOTAL P.W. OPERATING & INITIAL COST	81,730	130,368	
Uniform Annual Costs	4. Uniform Initial Cost UCR = .10185	1,283	873	
	5. Uniform Operating Cost UCR = .10185	8,324	12,405	
	6. Total	9,607	13,278	

CHAPTER 11

11-1 Depreciation: 1. 80,000 × 1/4 = 20,000
                    2. 60,000 × 1/4 = 15,000
                    3. 45,000 × 1/4 = 11,250
                    4. 33,750 × 1/4 = 8,438
                    5. 25,312 × 1/4 = 6,328 etc.

1. 20,000 (.6) + 20,000 (.4) = 20,000   20,000
2. 20,000 (.6) + 15,000 (.4) = 18,000   38,000
3. 20,000 (.6) + 11,250 (.4) = 16,500   54,500
4. 20,000 (.6) + 8,438 (.4) = 15,374   69,875
5. 20,000 (.6) + 6,328 (.4) = 14,531   84,406

Answer: 4.7 years

11-2

1. $20,000/1.10$ or $20,000 \times .90909 = 18,182$ 18,182
2. $18,000/1.10^2$ or $18,000 \times .82645 = 14,876$ 33,058

3. $16,500/1.10^3$ or $16,500 \times .75132 = 12,397$ 45,455
4. $15,375/1.10^4$ or $15,375 \times .68302 = 10,501$ 55,956
5. $14,531/1.10^5$ or $14,531 \times .62092 = 9,023$ 64,979
6. $13,899/1.10^6$ or $13,899 \times .56448 = 7,846$ 72,825
7. $13,424/1.10^7$ or $13,424 \times .51316 = 6,889$ 79,714

Answer: 7 years

11-3 $c = 38,000$ $s = 12,000$ $i = .10$

$$n = \frac{\log\left[\frac{1}{1-3.17(.10)}\right]}{\log 1.10} = \frac{\log 1.4641}{\log 1.10} = \frac{.16557}{.04139} = 4.0$$

Note: This problem could have been solved also by using the escalation log formula, letting $e = 0$.

11-4 The statement is correct provided it means that both the accumulated savings and the original investment are invested at the interest rate.

11-5 $60,000 \times .15 \times 3200 \times 1\frac{1}{4} \times .001 \times .03 = \$1,080$

$$60,000 = 1080 \left[\frac{1.07}{1.09} + \left(\frac{1.07}{1.09}\right)^2 + \ldots\ldots\right]$$

$$5.5556 = \left[\frac{1.07}{1.09} + \left(\frac{1.07}{1.09}\right)^2 + \ldots\ldots\right]$$

True Payback Chart: $\frac{1.07}{1.09} = .98$ and $\frac{c}{s} = 5.55$

Answer: Payback = 5.9 years
              By logarithms, the solution is 5.92 years.

11-6 Initial cost $\$20 \times 400 =$ $\$8,000$

Annual energy savings:
     $400 \times \$1.20 = \$480$
     P.W. of \$480 per year growing at 6%
     and discounted at 9% for 15 years is
     $480 \times$ (D.E.F., 15 years, .06/.09) =   (5,801)
     HUD grant =   (400)*
     Tax credit = $\$2,000 \times .30$ (600)
                     $\$6,000 \times .20$ (1,200)

Answer: Questionable investment   ($1)

---
*HUD grant must be included in gross income for tax purposes.

11-7 Annual energy cost of a motor is: KW of the motor × hours used in the year × cost per KWH divided by the efficiency rating of the motor.
Annual cost of energy efficient motor:

$$\frac{25 (.746) \times .04 \times h}{.914} + 739 \times (UCR; 30, 10\%)$$

Annual cost of standard motor:

$$\frac{25 (.746) \times .04 \times h}{.882} + 628 \times (UCR; 30, 10\%)$$

Break-even point is the number of hours at which these expressions will be equal. Set them equal and solve for h. Answer: h = 392.5 hours per year.

11-8 $40,000

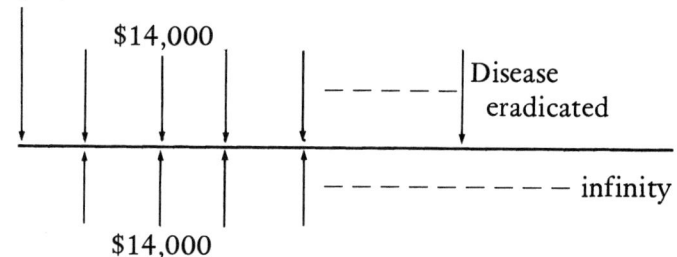

P.W. of costs = 40,000 + 14,000 × (UPW; Y, 10%)
P.W. of savings = 14,000 ÷ .10 = 140,000
40,000 + 14,000 (UPW; Y, 10%) = 140,000
        14,000 (UPW; Y, 10%) = 100,000
              UPW; Y, 10% = 7.14286

Look up the 10% compound interest factor page and go down the UPW column to the closest figure. Answer: 13+ years.

11-9 Let M represent the miles per year to break-even.
Cost of leasing: $4,400 × .50 annually for fixed costs and $.14 × M × .50 annually for mileage
i.e., 2,200 (UPW; 5, 12%) + .07 × M × (UPW; 5, 12%)
Cost of buying: $12,000 initial cost plus $.16 × M × .50 annually for mileage less the depreciation tax benefit of $1,000 a year less the salvage.

Solutions to Problems

i.e., 12,000 + .08 × M × (UPW; 5, 12%) − 1,000 (UPW; 5. 12%) − 2,000 (SPW; 5, 12%)
Equate the cost of leasing and the cost of buying and solve for M. Answer: M = 18,594 miles per year.

CHAPTER 11

12-1 P.W. of Initial Cost $150,000
P.W. of cost at end of year 1
50,000 × (SPW; 1, 7)
50,000 × .93458   =   46,729
Total P.W. of Investment   = $196,729
Present Worth of Savings
40,000 × (UPW; 10, 10) − 40,000 × (SPW; 1, 10)
40,000 × 6.14455 − 40,000 × .90909
245,782 − 36,364   = $209,425

$$\text{SIR} = \frac{\text{P.W. Saving}}{\text{P.W. Investment}}$$

$$\text{SIR} = \frac{209{,}425}{196{,}729} = 1.06$$

12-2 P.W. of Initial cost $150,000
P.W. of cost at end of year 1
50,000 × (SPW; 1, 10) × (SCA; 1, 5)
50,000 × .90909 × 1.05   =   47,727
Total P.W. of Investment   = $197,727
Present Worth of Savings
40,000 × (DEF; 10, 10, 8) − 40,000 × (SPW; 1, 10) × (SCA; 1, 8)
40,000 × 9.052914 − 40,000 × .90909 × 1.08
362,117 − 39,273   = $322,844

$$\text{SIR} = \frac{\text{P.W. Savings}}{\text{P.W. Investment}}$$

$$\text{SIR} = \frac{322{,}844}{197{,}727}$$

SIR = 1.63

12-3 Steam Line
Initial investment $ 20,000
P.W. of savings = 6,000 × (DEF; 10, 10, 8)
6,000 × 9.052914   =   54,317

$$\text{SIR} = \frac{\text{P.W. Savings}}{\text{Initial investment}}$$

$$\text{SIR} = \frac{\$54{,}317}{20{,}000} = 2.72$$

Shelter Installation
Initial investment $\quad\quad\quad\quad\quad\quad\quad\quad$ \$ 18,000
P.W. of savings = 5,000 × (DEF; 10, 10, 8)
$\quad\quad\quad\quad\quad\quad$ = 5,000 × 9.052914 $\quad$ = $\quad$ 45,265

$$\text{SIR} = \frac{\$45{,}265}{18{,}000}$$

SIR = 2.51

CHAPTER 13

13-1

```
1              INTEGER Y(40)
2              REAL C(40)
3     23       FORMAT (' ' ,7X, 'INITIAL', 6X, 'REPLMT', 7X,
              1 'ENERGY', 6X, 'OP&M', 7X, 'TOTAL', 5X, 'AV.ANN.',
              2 /, ' ', 'YRS', 6X, 'COST', 7X, 'COST', 9X, 'COST', 8X,
              3 'COST', 8X, 'COST', 8X, 'COST')
4     25       FORMAT (' ', I2, F12.0, F12.0, F12.0, F12.0, F12.0,
              1 F12.0)
5     26       FORMAT (' ', 'EQUIPMENT COST =', F10.0, 6X,
              1 'DISCOUNT RATE =', F10.2, /,
              2 ' ', 'STRUCTURE COST =' F10.0, 6X, 'CURRENT
              3 O&M COSTS =', F10.0, /,
              4 ' ', 'ENERGY ESC. RATE =', F10.3, 6X, 'CURRENT
              5 ENERGY COST =', F10.3, /,
              6 ' ', 'O&M ESC. RATE =', F10.3, 8X, 'ANNUAL KWH
              7 USAGE =', I10, /,
              8 ' ', 'REPLACEMENT ESCALATION RATE =', F10.3)
6     27       FORMAT (' ', 'REPLACEMENT COSTS:', 4X, 'YEAR',
              1 6X, 'AMOUNT')
7     29       FORMAT (24X, I2, F13.0)
8              READ (5, *) DR, RE, RO, RR
9              READ (5, *) N, NN
10             READ (5, *) AO, KW, CK
11             READ (5, *) EC, SC
12             DO 100 I = 1, 40
13    100      READ (5, *, END = 101) Y(I), C(I)
```

14		I = 41
15	101	II = I − 1
16		WRITE (6, 26) EC, DR, SC, AO, RE, CK, RO, KW, RR
17		IF (II. EQ.0) GO TO 105
18		WRITE (6, 27)
19		DO 28 I = 1, II
20	28	WRITE (6,29) Y(I), C(I)
21	105	WRITE (6,23)
	C LET	FCOST REPRESENT TOTAL FIRST COST.
22	22	FCOST = EC + SC
23		RCOST = 0
24		IF (II.EQ.0) GO TO 103
	C LET	RCOST BE PRESENT WORTH OF REPLACEMENT COSTS.
25		DO 102 I = 1, II
26		X = ((1 + RR) / (1 + DR)) ** Y (I)
27		IF (N.GE.Y (I))
	1	RCOST = RCOST + C (I) * X
28	102	CONTINUE
	C LET	ECOST BE THE PRESENT WORTH OF ALL ENERGY COSTS.
29	103	ECYR1 = KW * CK * (1 + RE * .6) / (1 + DR * .6)
30		E = (1 + RE) / (1 + DR)
31		ECOST = ECYR1 * ((1 − E ** N) / (1 − E))
	C LET	OCOST BE THE PRESENT WORTH OF OP. & MAINT. COSTS.
32		OCYR1 = AO * (1 + RO * .6) / (1 + DR * .6)
33		O = (1 + RO) / (1 + DR)
34		OCOST = OCYR1 * ((1 − O ** N) / (1 − O))
	C LET	TCOST BE THE PW OF ALL COSTS.
35		TCOST = FCOST + RCOST + ECOST + OCOST
	C LET	AVCST BE THE TCOST CONVERTED TO AV. ANNUAL COST.
36		UCR = (DR * ((1 + DR) ** N)) / ((1 + DR) ** N−1)
37		AVCST = TCOST * UCR
38		WRITE (6,25) N, FCOST, RCOST, ECOST, OCOST, TCOST, AVCST
39		IF (N.GE.NN) STOP
40		N = N + 1
41		GO TO 22
42		END

292   Life Cycle Costing: A Practical Guide for Energy Managers

EQUIPMENT COST = 9000000.   DISCOUNT RATE = 0.10
STRUCTURE COST = 9000000.   CURRENT O&M COSTS = 430000.
ENERGY ESC. RATE = 0.070   CURRENT ENERGY COST = 0.030
O&M ESC. RATE = 0.070   ANNUAL KWH USAGE = 9500000
REPLACEMENT ESCALATION RATE = 0.060
REPLACEMENT COSTS: YEAR   AMOUNT
                      15   9000000.
                      30   2000000.

YRS	INITIAL COST	REPLMT COST	ENERGY COST	OP&M COST	TOTAL COST	AV. ANN. COST
10	18000000.	0.	2481624.	3744204.	24225808.	3942652.
11	18000000.	0.	2694104.	4064789.	24758880.	3811966.
12	18000000.	0.	2900790.	4376630.	25277408.	3709806.
13	18000000.	0.	3101838.	4679967.	25781776.	3629530.
14	18000000.	0.	3297403.	4975030.	26272416.	3566391.
15	18000000.	5163450.	3487636.	5262047.	31913104.	4195745.
16	18000000.	5163450.	3672679.	5541235.	32377344.	4138372.
17	18000000.	5163450.	3852676.	5812810.	32828912.	4092597.
18	18000000.	5163450.	4027765.	6076979.	33268176.	4056404.
19	18000000.	5163450.	4198078.	6333942.	33695440.	4028191.
20	18000000.	5163450.	4363746.	6583897.	34111072.	4006680.
21	18000000.	5163450.	4524892.	6827031.	34515344.	3990823.
22	18000000.	5163450.	4681649.	7063541.	34908624.	3979767.
23	18000000.	5163450.	4834131.	7293601.	35291168.	3972795.
24	18000000.	5163450.	4982453.	7517385.	35663264.	3969318.
25	18000000.	5163450.	5126727.	7735062.	36025216.	3968831.
26	18000000.	5163450.	5267068.	7946804.	36377296.	3970913.
27	18000000.	5163450.	5403583.	8152774.	36719776.	3975198.
28	18000000.	5163450.	5536374.	8353126.	37052928.	3981379.
29	18000000.	5163450.	5665545.	8548015.	37376976.	3989177.
30	18000000.	5821751.	5791189.	8737583.	38350496.	4068195.

The above information suggests that in Year 14 any alternatives to the large replacement cost should be considered. At that time the input information can be updated for better projections. If the above information is still valid and the replacement made, the low point in average annual costs in Year 25 indicates that alternatives should be examined at that time.

Here is the same problem, but solved with an energy escalation rate of .09 rather than .07.

EQUIPMENT COST = 9000000.   DISCOUNT RATE = 1.10
STRUCTURE COST= 9000000.   CURRENT O&M COSTS = 430000.
ENERGY ESC. RATE = 0.090   CURRENT ENERGY COST = 0.030
O&M ESC. RATE = 0.070   ANNUAL KWH USAGE = 9500000
REPLACEMENT ESCALATION RATE = 0.060
REPLACEMENT COSTS: YEAR   AMOUNT
                     15   9000000.
                     30   2000000.

YRS	INITIAL COST	REPLMT COST	ENERGY COST	OP&M COST	TOTAL COST	AV. ANN. COST
10	18000000.	0.	2720702.	3744204.	24464880.	3981560.
11	18000000.	0.	2979354.	4064789.	25044128.	3855884.
12	18000000.	0.	3235655.	4376630.	25612272.	3758952.
13	18000000.	0.	3489626.	4679967.	26169568.	3684122.
14	18000000.	0.	3741288.	4975030.	26716304.	3626647.
15	18000000.	5163450.	3990663.	5262047.	32416128.	4261880.
16	18000000.	5163450.	4237770.	5541235.	32942432.	4210600.
17	18000000.	5163450.	4482630.	5812810.	33458864.	4171130.
18	18000000.	5163450.	4725265.	6076979.	33965680.	4141451.
19	18000000.	5163450.	4965691.	6333942.	34463056.	4119957.
20	18000000.	5163450.	5203934.	6583897.	34951248.	4105367.
21	18000000.	5163450.	5440015.	6827031.	35430464.	4096633.
22	18000000.	5163450.	5673942.	7063541.	35900912.	4092893.
23	18000000.	5163450.	5905746.	7293601.	36362784.	4093429.
24	18000000.	5163450.	6135445.	7517358.	36816256.	4097646.
25	18000000.	5163450.	6363054.	7735062.	37261536.	4105034.
26	18000000.	5163450.	6588593.	7946804.	37698832.	4115171.
27	18000000.	5163450.	6812082.	8152774.	38128288.	4127681.
28	18000000.	5163450.	7033538.	8353126.	38550096.	4142251.
29	18000000.	5163450.	7252984.	8548015.	38964416.	4158601.
30	18000000.	5821751.	7470432.	8737583.	40029744.	4246328.

Year 14 is again a critical year for reexamination of alternatives. Beyond that, Year 22 is the next low point in average annual costs and an appropriate time for another study.

CHAPTER 14

14-1 Step One:
    Debt   40,000,000 = 50%
    Equity 40,000,000 = 50%
              80,000,000

Step Two:
    Debt rate: 60% × 8% = 4.8%
    Equity rate: $\frac{4.80}{60}$ + 6% = 14.0%

	Weight ×	Rate =	Weighted Rate
Debt	.50	4.8	2.4
Equity	.50	14.0	7.0
Weighted average cost of capital		=	9.4%

14-2

	Amount	× Rate =	
New debt	$ 10,000,000	.06	$ 600,000
Old debt	40,000,000	.048	1,920,000
Equity	50,000,000	.14	7,000,000
	$100,000,000		$9,520,000

New annual cost $9,520,000
Old annual cost   7,520,000
                   $2,000,000

$\frac{2,000,000}{20,000,000}$ = 10% Marginal C.C.

14-3 0% Discount Rate

ABC —	Initial Cost		$115,000
	Annual Costs 20 × $8,000		160,000
	Total		275,000
DEF —	Initial Cost		100,000
	Annual Costs 20 × $9,000		180,000
	Total		280,000

7% Discount Rate
    ABC —   Initial Cost   115,000

	Annual Costs $8,000 (UPW, 20, 7%)	84,752
	Total	199,752
DEF —	Initial Cost	100,000
	Annual Costs $9,000 (UPW, 20, 7%)	95,346
	Total	195,346

DEF has the lower life cycle cost using the 7% discount rate. Comment: The critical question here is whether a 0% or a 7% discount rate is to be used. There is extensive literature available on the subject of government discount rates, and although there is disagreement on how the rate should be determined, there is no support for a 0% rate. As a matter of fact, even the 7% rate on the state's bonds is considered too low. Of course, at any rate higher than 7% DEF's equipment would be even more superior.

Index

Abbreviations, 69
Accelerated depreciation, 101
Additional first-year depreciation, 201
ADR, 103
After-tax cost of debt, 110
Air leakage cost, 80
Annuity, 22

Baseline year, 39
Break-even analysis, 130, 136
Building energy cost, 71
Btu method, 144

Capital budgeting, 12
Cash flow diagram, 15
CER, 55
Chain method, 115
Compounding, 13
  continuous, 27
  frequency of, 23
Computer analyses, 150
Continuous cash flow, 27
Costs, 49
  determination, 51
  estimating, 55
Cost of capital, 13

Demand cost, 84, 86
Department of Energy, 147
Depreciation, 50, 100
Discounting, 13
Discount-escalation, 125, 188
Discount rate, 110, 157
Double declining balance, 102

Electrical power cost, 84
Energy conservation credits, 198
Energy cost, 84
  estimating, 67
Energy investment credit, 200
Energy savings, 67

Energy values, 70
Escalation, 119
  present worth method, 123
  average annual cost method, 127
Expert judgment method, 56
Expected value analyses, 62
Extrapolation method, 57

Federal Energy Management Program, 144
Forever project, 45
Format for LCC, 51, 52
FORTRAN IV, 150
Fuel adjustment cost, 84
Future value, 15

Heat loss, 72

Infiltration air cost, 76
Information sources, 54
Interest cost, 49
Interest rate, 15
  effective rate, 23
  nominal rate, 23
  number of interest periods, 15
  real rate, 162
Interest tables, 22
Interval cost, 125
Investment rates, 162
Investment tax credit, 108

LCPBM, 153
Lease, 107
  leverage lease, 108
Life Cycle Costing (LCC)
  current use, 2
  definition, 1
  objective, 2
  opportunity, 6
  where to use, 3
Life cycle planning and budgeting model, 153

Index

Lighting cost, 71
Logarithm method, 190
Low bid, 11

Marginal cost of capital, 160
Make-up air cost, 79
Motor operating cost, 82
MTTR, 57

Office of Management and Budget, 163
Open door infiltration air cost, 91
Opportunity cost, 13, 50

Parameters, 56
Parametric costing, 55
Payback, 130, 142
  discounted, 131
  logarithm method, 134
  simple, 130
  true, 134
Pollution control facilities, 201
Power factor, 84
Present worth (value), 13, 15, 37, 39
Public sector discount rates, 163
Pump cost, 83

Ranking, 141
Regular investment credit, 199
Renewable energy source credits, 198
Replacements, 112
Risk adjustment, 161

Savings/investment ratio, 142
SAS, 59, 60
Sensitivity analysis, 61, 153
Service life, 93
Single Compound Amount (SCA) 15
Single Present Worth (SPW), 17
SIR, 142
Solar applications, 183
  taxes, 198
Straight-line depreciation, 100
Steam line cost, 87
Steam leaks, 88
Steam trap leaks, 90
Sum-of-the-years digits, 102

Tax consideration, 99
Time value of money, 11, 12, 13

Uniform annual cost, 39, 42
Uniform Capital Recovery (UCR), 18
Uniform Compound Amount (UCA), 21
Uniform Present Worth (UPW), 19
Uniform Sinking Fund (USF), 20
Uniform sum of money, 15
UNIFORMAT, 2, 153

Variables, 56

Water heating cost, 78
Weighted Average Cost of Capital (WACC), 158
Window transmission heat loss, 78

NOTES

NOTES

NOTES

NOTES

NOTES

NOTES

NOTES

NOTES

NOTES